Gulf of California Coastal Ecology: Insights from the Present and Patterns from the Past

Markes E. Johnson

Jorge Ledesma-Vázquez

SUNBELT PUBLICATIONS

San Diego, California

Gulf of California Coastal Ecology: Insights from the Present and Patterns from the Past

Sunbelt Publications, Inc.
Copyright © 2016 by Markes E. Johnson and Jorge Ledesma-Vázquez
All rights reserved. First edition 2016

Edited by B.Gudveig Baarli
Cover and book design by B. Gudveig Baarli
Project management by Debi Young
Printed in the United States of America

Sunbelt Publications, Inc.
P.O. Box 191126
San Diego, CA 92159-1126
(619) 258-4911, fax: (619) 258-4916
www.sunbeltbooks.com

19 18 17 16   4 3 2 1

Library of Congress Cataloging-in-Publication Data

Johnson, Markes E., author.
  Gulf of California coastal ecology : insights from the present and
patterns from the past / Markes E. Johnson & Jorge Ledesma-Vazquez. --
First edition.
        pages cm
  Includes bibliographical references.
  ISBN 978-1-941384-18-3 (softcover : alk. paper) 1. Coastal ecology-
-Mexico--California, Gulf of. I. Ledesma-Vazquez, Jorge, author. II.
Title.
  QH107.J63 2015
  577.5'1--dc23
                                                        2015031313

All photographs are by the authors unless noted or in public domain.

Dedication

Geographically and culturally, the gulf region that surrounds the Gulf of California on the western frontier of Mexico is unique. Compared to the rest of Mexico and most other parts of the world, population centers remain relatively small and the Baja California peninsula continues to be home to a hardy people accustomed to enjoying the fruits of an inland sea bordering on a dessert land. The missions of this booklet are 1) to describe the riches with which nature has endowed the region, 2) to instruct how we can better understand its ecological complexities in the context of its geological heritage, and 3) to promote the ways we must be proactive to care for such a great treasure. The gulf region shares with other attractive regions all around the world the challenge of controlling progress in such a way that the natural environment is protected from growing human pressure. While the Mexican federal government has put into place measures to provide for marine bioreserves and marine parks, the public must be informed and vigilant in order to maintain and even expand protections for one of the world's most precious and instructive windows on ecology. This booklet is thus dedicated to the children and young adults in whose care the future of the Gulf of California rests.

Contents

Illustrations

Figures

vi

cepción peninsula.

Plates

1. Top: Isla San Luis (left), Cabo San Lucas arch (right); Middle: San Basilio rhyolite (left), El Mangle limestone (right); Bottom: Isla Coronados basalt (left), Isla Monserrat andesite agglomerate (right).

2 Top: Mulegé dune with wind etching (left), San Nicolás dunes (right); Middle: El Rinchón shell beach (left), Monserrat shell beach (right); Bottom: Punta Chivato and Islas Santa Inés (left), Isla Coronados (right).

3. Top: Big lagoon Isla Angle de la Guarda (left), small lagoon (arrow) Isla San Lorenzo (right). Middle: Bacterial filaments, 10 microns across (left), solitary bacterium, 30 microns (right); Bottom: Punta Arena de la Ventana salt pond (left), bacterial dome (right).

4. Top: Isla Cerralvo fan deltas, numbered 1-6 (west shore); Upper middle: Killer whales off south end of Isla del Carmen; Lower middle: Sea lions on Las Galeras; Bottom: Pliocene whale-bone fragments from Mesa Ensenada de Muerte.

Vocabulary list

Acknowledgements

Inspiration for this project came from a handbook for high school students living in the Caribbean islands written by biologists Alick Jones and Nancy Sefton (*Marine Life of the Caribbean*, 1978, Macmillan Caribbean). Although appropriate for eco-tourists, what is most impressive about the handbook is the care taken by scientists to promote a better understanding of coastal ecosystems for students growing up in the very places where the ecosystems are located. Healthy ecosystems require that local residents have the knowledge they need to understand and take pride in the natural resources from their own back yard.

The authors owe a huge debt of gratitude to B. Gudveig Baarli, who took on the task of designing this book. She did this out of an interest in the project and her own life-long enthusiasm for exploring all aspects of nature.

As scientists and teachers, we have worked together for more than 25 years, bringing students from Williams College and the Universidad Autónoma de Baja California on field trips to the Gulf of California. Students often have their own way of seeing things that cast a new light on subjects their professors may fail to see. Johnson is especially grateful to former students Marshall Hayes, Max Simian, Jon Payne, Mike Eros, and Peter Tierney for their fine company and their insights during field studies. Likewise, Ledesma-Vázquez acknowledges former students Itzel Hernandez-Morlan, Francisca Staines-Urias, Yan Ruiz-Taboada, and Astrid Montiel-Boehringer.

Leon Fichman and Ivette Granados (Loreto), and Tim Hatler (La Ventana), rendered unfailing services as outfitters for our island adventures. Mexican government offices issuing permits that made research possible in national parks and biosphere reserves included the Área de Protección de Flora y Fauna Islas de Golfo de California, Comisión de Áreas Naturales Protegidas (CANP); Bahía Parque Nacional (BLPN); and the Instituto Nacional de Antrropoligía e Historia (INAH) en Baja California Sur. Alfredo Zavala (CANP) and Everardo Mariano-Meléndez are due our thanks for their cooperation and interest in our research.

Norm Christie (part-time resident of Loreto) deserves credit for agreeing to read early drafts of the book, correcting mistakes and numerous typos. Tom Bowen is a valued correspondent who shared information regarding microbial mats in the northern gulf region and his research on Federico Craveri. Rafael Riosmena-Rodríguez has been a steady source of encouragement for this project. It is a pleasure to thank Debi Young, procurement editor at Sunbelt Publications, for her enthusiasm in seeing the project through to completion. The cover photo for this book is reprinted with permission from the January 2012 cover photo from the *Journal of Coastal Research* from an original image taken by MEJ. Finally, the authors are grateful for financial support through the Office of the Dean of Faculty at Williams College, Williamstown, Massachussetts that made publication feasible.

Chapter 1

Classroom for Ecology in an Over-crowded World

Introduction

Few other regions the same size outside Mexico's Gulf of California offer the opportunity to experience and explore the intricate web of nature that binds together so many different coastal ecosystems on such a dynamic scale. Recognized as one of the planet's ecological hot spots for marine life due to high levels of productivity, the gulf's coastal zones envelop many islands and embayments large and small that serve as crucial nurseries for organisms. It is easy to become enamored with the big players – the whales, dolphins, sea turtles, and other vertebrates that thrive on the sea's bounty. The Gulf of California also hosts an amazing array of invertebrate and algal life with natural-history cycles that play a profound role in the replenishment of beaches and coastal sand dunes. Organisms that secrete calcium carbonate ($CaCO_3$) include invertebrates like corals and mollusks (clams and gastropods, among others), as well as coralline red algae that form spherical growths capable of rolling around on the seabed.

These entities occupy their own geographic territories defined by boundaries as real as those for nation states. The neighborhoods where corals congregate are different from those where clams tend to prosper, or where the bizarre rhodoliths (rolling coralline algae) form major banks. The region's climate, which follows seasonal patterns of changing wind and water circulation, constitutes the physical enforcement that maintains these biological boundaries. That is to say, many organisms react to the power of winds and waves differently according to their threshold for engagement with the physical environment. Other distinct habitats include the mangroves that grow as salt-tolerant plants on the fringes of open lagoons and simple microbes that form thick bacterial mats in closed lagoons.

Thus, the organization of life in distinct settings within the Gulf of California is truly broad, ranging from lowly bacterial forms to socially adept feeders like dolphins and whales at the apex of the system as primary consumers. The gulf's coastal ecology is so fulsome that aspects of its nature may be observed almost anywhere that short road trips or boat rides may bring an inspired explorer. However, this is only half the story that the Gulf of California has in store for those who would look further. The gulf's geological history records multiple interlocking ecosystems that date back several million years in age. Sedimentary rock layers, often dominated by fossil-rich limestone, are exposed along the shores where the very same organisms preserved as fossils may be found today living in relatively shallow waters. These rock formations testify to the prehistoric richness of the Gulf of California, and they serve as geological monuments to the past. They also provide an important record against which the health and vitality of various gulf ecosystems may be measured.

Meaning of Ecology

Ecology is the formal name for the study of organisms based on how they interact with one another in loose associations with little dependence on other species or as members of communities with closer connections to other species for mutual survival. Either way, organisms that inhabit a particular environment must be able to tolerate variables in physical conditions such as temperature, wind intensity, wave agitation, water salinity, and water clarity. **Paleoecology** is the formal name for the study of past life as represented by fossils, but with much the same attention paid to the reconstruction of former environments with regard to the impact of abiotic factors. The key root shared in these terms is "ecos" – a transliteration from the Greek word "oikos" for house. The place we inhabit is our house, a setting with definite physical limitations in which we may live a relatively independent life or one more fully engaged with other species around us. A closely related concept is that of the ecosystem, in which the basic natural history of a specific environment together with its resident life forms may be studied in the context of how whole populations interact with the populations of other species in terms of nutrient cycles and energy flow from self-nourishing producers (marine algae and other plants) to primary and secondary consumers (most marine invertebrates and all vertebrate animals).

It is mainly during the last 75 years that the gulf's coastal ecosystems have come under scrutiny by explorers and research specialists. Marine biologist Ed Ricketts (1897–1948) together with writer John Steinbeck (1902-1968) came to Baja California aboard the chartered fishing vessel *Western Flyer* in 1940 in a legendary collaboration celebrating the holistic outlook that everything in nature is somehow interconnected. First published in 1941 under the title *Sea of Cortez: A Leisurely Journal of Travel and Research*, their account fails to mention the word ecology. With a central focus on intertidal rocky shores and sand flats, however, the contribution marks the beginning of modern marine ecology in the Gulf of California. Later in 1940, another group of scientists reached the Gulf of California aboard the research schooner *E.W. Scripps*. Among the crew were geologist Charles A. Anderson (1902-1990), paleontologist J. Wyatt Durham (1907-1996), and oceanographer Francis P. Shepard (1897-1985). The results of their work were not issued until 1950 as Memoir 43 of the Geological Society of America, but their combined reports set the baseline for modern paleoecology in the Gulf of California.

Birth of a Young Sea

A valuable outcome of the *E.W. Scripps* expedition was production of a set of bathymetric charts for the Gulf of California coordinated by F.P. Shepard. A series of 11 detailed maps for smaller areas went into the compilation of a single, large chart (46 cm

by 120 cm) covering the entire gulf. For the first time with publication in 1950, it was possible to see the configuration of several deep-water basins that subdivide the region. Today with input by geophysicists, these segmented basins are understood as part of the tectonic framework within the Gulf of California (Fig. 1.1), represented by multiple centers for seafloor spreading offset by a succession of transform faults. Reaching across the international border from Mexico into southern California, this system is linked to the famous San Andreas Fault that threads its way as far north as San Francisco and beyond to exit into the Pacific Ocean. The entire Baja California peninsula together with the western margin of Alta California is shifting to the northwest, while lands on the opposite side of the fault network are slowly moving to the southeast. Opposing arrows in the accompanying tectonic map show the direction of crustal movement on the seafloor.

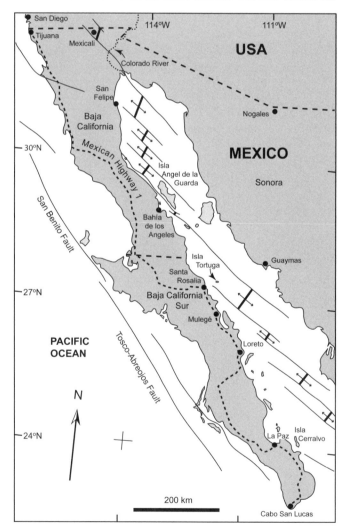

Fig. 1.1 Tectonic map for the Gulf of Canifornia.

Prior to about 12 million years ago, the Gulf of California did not exist and what is now the Baja California peninsula was attached to mainland Mexico. Up until that time, dense oceanic crust (represented by basalt) off the west coast of Mexico pushed uniformly against and beneath the lighter crust of the North American continent (represented by granite) in a process called **subduction**. Separation of the peninsula from the Mexican mainland began when the continental crust became stretched through a process called extensional rifting. This process continued for some 8.5 million years until about 3.5 million years ago as the protogulf slowly opened in an east-west fashion (Ledesma-Vázquez et al., 2009). Rifting is associated with the initial subduction of the East Pacific Rise (main locus of rifting on the floor of the Pacific Ocean) beneath the North American continent. Many islands in the western Gulf of California were raised as fault blocks (called **horsts**), when normal (up and down) faulting occurred during this stretching process. Passages between the mainland and the islands formed when neighboring fault blocks (called **grabens** or half-grabens) sank into the thinning crust to form depressions that later filled with seawater.

The present-day regime of transtensional tectonics took hold about 3.5 million years ago with the shift to lateral motion on opposite sides of transform faults fractured perpendicular to the spreading centers captured within the protogulf. Progress in the northwesterly migration of the peninsula may be visualized by gauging the amount of offset between neighboring segments of the spreading centers. For example, the offset between the adjacent spreading centers located between Guaymas and Bahía de los Angeles amounts to roughly 250 km (Fig. 1.1). Only a few million years ago the spots now occupied by those two towns were situated across from one another on opposite shores of the Gulf of California. From Cabo San Lucas in the south to San Francisco in the north, subduction no longer impacts this part of continental North America, although the traces of former activity remain off shore represented by the submarine San Benito and Tosco-Abreojos fault scarps. In terms of global tectonics, Baja California and the western margin of Alta California no longer belong to the North American Plate, but are transferred to the Pacific Plate.

Wind and Marine Circulation

Atmospheric forcing directly stimulates wind, wave, and surf conditions throughout the Gulf of California. The region's wind fields are seasonal and semi-monsoonal in character (Bray and Robles, 1991). During the colder winter months from November to April, high-pressure systems rotate in a clockwise fashion above the southwest deserts of North America and spin off cells that are pulled southward toward the equator through the gulf's axis. These winter winds may blow for several days at a time. Due to long distances over open water (called **fetch**), these south-directed winds generate broad wave trains that result in strong surf against north-facing peninsular headlands and the north-facing shores of islands. When the land heats up during the summer months from May through October, lighter winds tend to flow from south to north through the gulf's axis. Automated weather stations strategically sited throughout the Gulf of California record this seasonal change in wind directions. One such station occupies a central location on Isla Tortuga between Santa Rosalía and Guaymas (Fig. 1.1). There, wind velocity may reach a peak of about 10 m/s during intervals of sustained activity in January. By comparison for the months of June and July, winds reaching Isla Tortuga from the south register a maximum velocity of 5 m/s, and typically much less.

Thermohaline circulation is another phenomenon influenced by the physical geography peculiar to the Gulf of California. More so in the far north, aridity is such that almost a meter of seawater is evaporated each year from the surface of the gulf (Bray, 1988). Were it completely enclosed and isolated from the open ocean, the Gulf of California would become a dry seabed and reveal a deep valley exposed well below sea level. However, the seawater evaporated farther north is constantly replenished by fresh seawater that inters the gulf from its opening on the Pacific Ocean to the south. Surface currents exiting the Gulf of California into the Pacific Ocean tend to carry seawater that is slightly elevated in salinity, while deep currents entering the gulf from the ocean bring colder seawater with normal salinity. Over the winter season, circulation patterns are counterclockwise with normal-salinity water flowing northward close to mainland Mexico, whereas more highly saline waters exit the gulf in currents near peninsular shores. The pattern is reversed to a clockwise flow during the summer months, when the salinity gradient hits a seasonal peak. Because the colder seawater drawn into the gulf as replacement water also carries a rich supply of nutrients that stimulates the growth of **phytoplankton** at the base of the **food chain**, the overall fertility of the region is greatly enhanced.

Tidal mixing is a third factor that has a strong bearing on marine circulation, especially in the far north from San Felipe to the delta of the Colorado River, where the tidal range is commonly on the order of 5 m. The effect of strong tides and tidal currents is to more thoroughly mix the seawater and its nutrients from top to bottom. During their 1940 visit, Ricketts and Steinbeck were impressed by the tidal range at the north end of Isla Angel de la Guarda, which they recorded as nearly 4 m at Puerto Refugio. Inflow of nutrient-rich water from the Colorado River and the growth of the Colorado River delta now severely restricted by upriver dams (Carriguiry and Sánchez, 1999), but formerly the river played a significant role in adding to the gulf's fertility.

El Niño Southern Oscillation

On an expanded regional scale covering the adjacent Pacific Ocean off Mexico and Central America, present-day El Niño Southern Oscillation (ENSO) cycles entail a warm phase linked with weakened trade winds that allow for the stagnation of warmer surface waters near the equator. The cool phase, or La Niña part of the cycle, is driven by more effective trade winds that converge on the equator to send cooler waters with enhanced upwelling westward

from the entrained California and Humboldt coastal currents. For western Mexico including all of Baja California, the ENSO warm phase signals the arrival of wetter winters. As the seasons turn to summer under amplified El Niño conditions, tropical storms that form off the Pacific shores of Mexico south of Acapulco are less likely to be diverted to the northwest (Romero-Vadillo et al., 2007). As a result, the frequency of hurricanes approaching the Baja California peninsula and entering the Gulf of California is increased during the warmest ENSO phases. According to the Oceanic Niño Index (ONI), strong El Niño years have a recurrence every five to seven years and this mirrors the frequency with which hurricanes are more likely to cause severe flooding in the southern Cape Region of Baja California.

The general pattern of regular turnover in ENSO phases observed historically during the past several decades could be typical for hundreds of thousands of years throughout many of the Pleistocene glacial and inter-glacial epochs. Based on data from deep-sea cores that reach farther back millions of years through the preceding Pliocene Epoch (about 2.5 to 5 million years ago), reconstruction of sea-surface temperatures and the depth of the **thermocline** along the equatorial Pacific Ocean suggest that a wet, El Niño-like climate was more common during what is called the **Pliocene Warm Period** (Wara et al., 2005). If correct, the interpretations point to a hurricane-prone environment that extended farther north into the Pliocene Gulf of California on a more regular basis than today. One of the goals of this handbook is to explore such a scenario for the Pliocene Gulf of California in contrast to the pattern of recent times based on clues preserved in region's rock record. This approach to paleoecology is relevant to concerns over global warming trends experienced today.

Kinds of Coastal Ecosystems

Rocky Shores. This environment with its attendant rocky-shore biota is most evident in places were the waves and surf are especially strong and marine organisms are adapted for firm attachment to intertidal rocks by virtue of strong ce-

Fig. 1.2 The granite arch at Cabo San Lucas.

Fig. 1.3. Fossil finger-coral colony from the Pleistocene of Isla del Carmen.

ments (as in the case of barnacles) or strong muscles (as in the case of chitons). The iconic rocky-shore image for all of Baja California is that of the great arch at the southern tip of the peninsula at Cabo San Lucas (Fig. 1.2). The arch is being eroded from solid granite by vigorous currents and surf that exfoliate the exposed rock surface.

Coral Beds and Reefs. The best example of a living coral reef is found off Cabo Pulmo in the southern Cape Region. Conditions limited by water temperature, water clarity, and the amount of available wave agitation determine where coral reefs may thrive. Today, lone coral colonies range far to the north in shallow coastal waters, but conditions are not suitable for those colonies to coalesce into fixed reef structures. When global temperatures were higher than today during the last interglacial episode some 125,000 years ago, coral reefs were more widespread through the gulf region. Colonies of large, branching corals, such as the common finger coral (*Porites panamensis*) became interlocked in large reef structures (Fig. 1.3) growing in places such as the south shore of Isla del Carmen.

Clam Flats. One of the most sought after clams (a molluscan bivalve) for human consumption is the chocolate Shell (*Megapitaria squalida*), so named

due to the creamy beige and brown coloration of its shell (Fig. 1.4). This species lives in large numbers beneath the sediment as an infaunal dweller in shallow sand flats. Hidden from view, the clams respire by the aid of a fleshy siphon that reaches up to the water-sediment interface. Where wave and tidal action is vigorous, the shells of many different bivalves and gastropods are washed onto beaches. Some of these mollusks are epifaunal dwellers that live exposed on top of the seabed.

Fig. 1.4. Modern chocolate shells from El Rincon.

Major clam flats occur along the San Marcos Passage, for example, between Isla San Marcos and the mainland peninsula near El Rinchón where the chocolate shell is a beachcomber's common prize. In addition to its culinary value, the chocolate shell is a major contributor to the enrichment of beaches and sand dunes along the gulf shores.

Rhodolith Banks. Some species of coralline red algae grow concentrically around a pebble or a shell fragment and remain unattached on the sea floor, adapting a free-rolling spherical or semi-spherical shape. **Rhodolith** (meaning "red stone") is the name applied to this sort of biological concretion. In life, rhodoliths are dark red to rose colored. Vast banks of rhodoliths accumulate throughout the Gulf of California in water as shallow as about 2 m. A good example of a rhodolith bank is found between Isla Coronados and the peninsular mainland at Punta Bajo, in this case dominated largely by a single species (*Lithothamnion margaritae*). Major storms sometimes sweep a multitude of rhodoliths onto the shore to be stranded in a supra-tidal setting, where they quickly bleach under the sun and expire. Individual rhodolith colonies are typically 5 cm in diameter, although the largest exceed 20 cm in diameter (Fig. 1.5) as shown by samples from Punta Bajo. Fossil rhodoliths commonly contribute to the fabric of Pleistocene and Pliocene limestone deposits throughout the Gulf of California.

Fig. 1.6. Beach and tombolo at El Requesón.

which the finer fraction of beach sand is blown inland to make coastal dunes. The popular camping place at El Requesón south of Mulegé in Bahía Concepción features a beach and tombolo that connects the mainland to Isla Requesón at low tide (Fig. 1.6). Healthy rhodolith banks are located off the north and south ends of Isla Requesón and the beach is dominated approximately 80% by the debris of wave-crushed rhodoliths. Beaches and related dunes can be enriched by the finely fragmented debris of clams and other mollusks, as found in the great dune field south of Mulegé (Fig. 1.7). In the southern Cape Region, beaches generally retain

Fig. 1.5. Modern rhodoliths from El Bajo.

Sandy Beaches and Coastal Dunes. The tides and waves that sweep clam flats and rhodolith banks are the means by which biological carbonates are shifted landward to become a significant component of beaches. Furthermore, prevailing winter winds are capable of causing beach deflation by

Fig. 1.7. Great sand dunes south of Mulegé.

little or no carbonate materials and the abiotic sand is concentrated from quartz grains eroded from granite. Limestone beach and dune deposits also are part of the Pleistocene record in Baja California.

Mangroves (Open Lagoons). Mangroves are characterized by different species of flowering plants that tolerate an inter-tidal setting in seawater or on the landward fringe of seawater. The plants grow as bushy thickets or even trees on or immediately behind the upper tide line, but prefer a setting in open lagoons protected from direct wave action. The tombolo at El Requesón joins Isla Requesón where concentrations of the red mangrove (*Rhizophora mangle*) have colonized the backside of the island (Fig. 1.6). Mangroves provide shelter for many kinds of invertebrates and fish during their juvenile stages of development, and the diversity of birds tends to be higher around the mangrove habitat. Fossilized mangrove wood is preserved at former mineral springs in Baja California where Pliocene rock formations have a high concentration of precipitated silica.

A World of Microbes (Closed Lagoons). Under constructive conditions of wave action and long-shore currents, what starts off as an open lagoon may evolve into a closed lagoon as migrating sand and cobbles shut off access with a fortified berm. Many such closed lagoons occur on gulf islands and along peninsular shores (Fig. 1.8). In recent years, some closed lagoons have been found to contain mat-like growths of microbial bacteria and blue-green algae.

Fig. 1.8. Closed lagoon with microbialites at El Quemado east of Bahía de los Angeles.

Microbial mats and domes known from the fossil record are called **stromatolites** (meaning "stoney layers"). The first extensive microbial mats from the Gulf of California were found in the closed lagoons on Isla Angel de la Guarda (Johnson et al., 2012), and more recently Pleistocene stromatolites have been identified from a former lagoon at Punta Chivato. Stromatolites represent the oldest known forms of life on Earth, and with the capacity for photosynthesis, they were responsible for slowly raising the level of oxygen on our planet.

Estuaries and Deltas. Flowing rivers and estuaries with deltas under continuous construction are a rarity in the parched desert lands of Baja California. A notable exception is the spring-fed Mulegé River that empties into the Gulf of California at Mulegé. However, the action of moving water is widely evident by the extensive development of arroyos that mark stream courses under flood conditions. Often, the largest arroyos join the Gulf of California at incised estuaries where seawater penetrates inland for a short distance. Similar to the mangrove habitat, such estuaries that rarely contain flowing water still provide a rich habitat for birds and for marine life. As discussed (above) under the heading of ENSO cycles, massive rock formations dominated by Pliocene conglomerate reflect an earlier time when hurricanes had a greater impact on the Baja California peninsula and the development of large fan deltas was more common than at present. Smaller deltas are still under construction, today, particularly in the southern Cape Region where hurricanes are more frequent than for the rest of the peninsula.

Coastal Hydrothermal Springs. Directly related to ongoing tectonic activity in the Gulf of California and associated with **fault systems**, hydrothermal springs are a part of the landscape. Some active springs are affected by the tides, such as the example encircled by mangroves at Santispac within Bahía Concepción. Other hydrothermal springs located in Bahía Concepción and farther south near Agua Verde are strictly subtidal in operation. Growing evidence indicates that the subtidal springs are sites where biodiversity is enhanced. Former hydrothermal springs often are identified with zones of mineralization on land and they represent activity that dates back to Pliocene time millions of years ago.

Denizens of the Open Sea. Whales, dolphins, sea turtles, rays, sharks, and a rich diversity of fish species belong to the open sea, but their activities often may be spotted from land. Whales breach and spout columns of water. Dolphins enjoy shadowing boats. Rays perform acrobatic leaps and somersaults. Perhaps the most common phenomenon easily observed from shore is that of a "fish boil" when a mass of smaller fish are frightened to the surface by larger predatory fish like Yellowtails. This particular activity is commonly viewed from the busiest harbor areas in San Felipe, Bahía de los Angeles, Santa Rosalía, Loreto, and La Paz. Paleoindians living along gulf shores were astute observers, who often recorded images of marine creatures in their rock art (Fig. 1.9) The populations of vertebrate animals speak to the great biological richness supported by the Gulf of California, particularly the many links in the regional food web. The region's spectacular fossil record also reminds us that whales, sea turtles, and rays arrived with the expanding Gulf of California as far back as 5 million years ago.

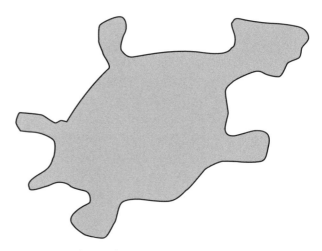

Fig. 1.9. Paleoindian rock art from the Concepción peninsula showing a sea turtle (size 40 cm x 40 cm).

Chapter 2

Rocky Shores

Introduction

In a survey of the Gulf of California's west shores and all its related islands, the combined coastal length measures 2,940 km based on satellite imagery (Backus et al., 2009). The breakdown between rocky shores and gravel to sandy shores is 48% to 52%, respectively. As the Gulf of California is part of a tectonically active zone where coastal uplift continues to be significant, its towering sea cliffs have a dramatic presence. The rocky shores that form such an imposing part of the region's landscape include all three principal rock groups: **igneous**, **metamorphic**, and **sedimentary** rocks. Among these, the igneous rocks dominate to account for 80% of all rocky shores in the western Gulf of California. Approximately 15% of gulf shores are formed by limestone. Metamorphic rocks represent the rest with about 5% of the total. Fundamentally, the erosion of rocky shores is constant due to wind-driven waves, strong currents, and attendant tidal action (particularly in the north) through much of the year. Thus, rocky shores are the focus of mechanical wear that exports sediments to beaches and bays in more protected settings.

Different kinds of specialized marine organisms colonize high-energy rocky shores. Chief among them is direct cementation to the rock surface, for example by barnacles, oysters, and corals. Other marine invertebrates, such as the chitons, cling to the rocks by means of strong muscles. Some species of clams (bivalves) hold onto rock surfaces by means of strong organic threads that act as attachment cables. These include the ark shells and the mussels. Other clams resist wave activity by wedging into rock crevices or fitting in between boulders at the base of sea cliffs. These include some bittersweet shells, some venus shells, and even some lucinid shells. Under conditions of rising sea level that promote onlap of marine sediments (also called a **marine transgression**), a former sea cliff may be buried and preserved in the geologic record together with many of the rocky-shore organisms that become fossils. Otherwise, the colonizing organisms are subject to erosion together with the rocks on which they lived, and may be removed by waves and currents to be deposited as shelly debris in neighboring bays or inlets.

Relevance of the Rock Cycle

Physical changes in coastal settings are apparent after major storms, perhaps every few years. Molten lava that issues from volcanoes and the rapid process through which lava cools and hardens into basalt or andesite also can be observed directly at the surface. Other processes operate over far longer periods in the context of geologic time. The concept of the **rock cycle** (Fig. 2.1) brings into focus several

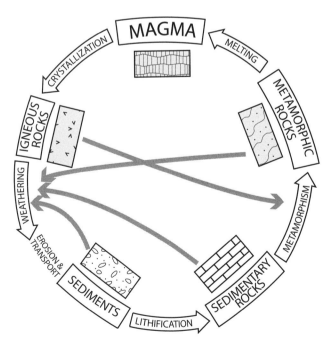

Fig. 2.1. The rock cycle relevant to igneous, sedimentary, and metamorphic rocks.

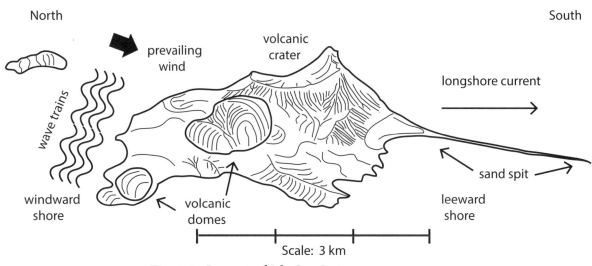

prevailing
wind

volcanic
crater

longshore current

wave trains

windward
shore

volcanic
domes

sand spit

leeward
shore

Scale: 3 km

Fig. 2.2. Layout of Isla San Luis.

other processes beyond that of volcanic eruptions or coastal attrition that are difficult to observe on a human time scale. Magma that fails to reach the Earth's surface, for example, requires a much longer time to crystallize into igneous rocks such as gabbro or granite. Additional time is required for these deeply buried rocks to be uncovered and exposed at the surface, where weathering and erosion first have the opportunity to generate sediments. Another slow process that occurs deep within the Earth is that of metamorphism, by which an igneous rock such as granite may be transformed by heat and pressure into a metamorphic rock like schist. Likewise, a sedimentary rock such as limestone may be changed into marble. Once again, deeply buried metamorphic rocks are not subject to weathering and the production of sediments until they become exposed at the Earth's surface.

The relationship is more intuitive between sediments that originate from the physical breakdown of igneous or metamorphic rocks due to weathering and the formation of sedimentary rocks like sandstone due to the process of lithification (Fig. 2.1). We may recognize the different components in a handful of beach sand composed of silica sand mixed with other minerals commonly found in granite such as plagioclase and biotite. In the same way, it is not difficult to see the similarities between a sample of beach sand and nearby sandstone that includes the identical components cemented together in solid form. Likewise, it is entirely cred-

ible to understand that a limestone deposit may be dominated by fragments of broken corals, or perhaps even formed by an entire reef structure preserved intact due to sudden, catastrophic burial. In this instance, the rock cycle may be considered to invert itself with the understanding that reef limestone exposed at the seashore might be repeatedly attacked by storms that free whole fossil corals. Subsequently, the fossils may be broken into pieces by the force of the waves to generate a kind of loose, coral gravel in the intertidal zone.

The Gulf of California is an extraordinary natural laboratory in which to visualize the full nature of the rock cycle. Separated by a distance of more than 900 km at opposite ends of the gulf, two examples serve to illustrate the dynamics of the rock cycle under local conditions. To the north, Isla San Luis is a small volcanic island (4.5 km^2 in area,) close to shore about 115 km south of San Felipe. The half-eroded remains of a large volcanic cone and crater rim that rise 180 m above sea level, as well as two smaller rhyolite domes nearby reveal the island's origins (Fig. 2.2). An islet located about 1 km north of Isla San Luis also shows signs of a related nature represented by the sunken section of an eroded crater rim. On one hand, the recent emergence of the island complex is shown by the freshness of the volcanic features. On the other hand, the island's ongoing erosion by prevailing winds and wind-driven waves out of the north makes a strong impression. From a sedimentologi-

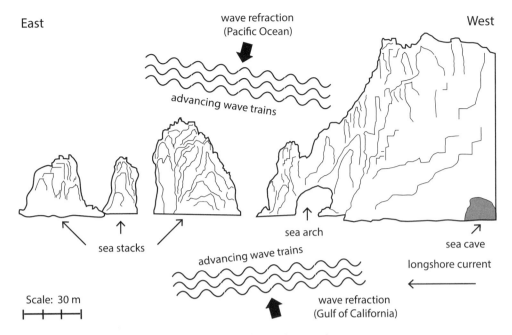

Fig. 2.3. Layout of Land's End at Cabo San Lucas.

cal standpoint, the enormous sand spit that extends for 1.25 km off the south end puts the power of the wind and associated long shore currents into context as erosive factors. The aerial view from which the drawing in Fig. 2.2 was created is shown in Plate 1, as photographed during an over-flight from an altitude of about 9,000 m. In that photo, the intensity of surf striking the windward side of the island is in contrast to the calm leeward side of the island, where sediment eroded from the shore reaches quieter water and settles out to form the spit.

To the south at Cabo San Lucas, the prominent spur of an eastward-directed ridge forms the famous landmark at land's end represented by a rock arch and sea stacks (Fig. 2.3). There, persistent wave trains join together from opposite directions at the end of the peninsula on the Pacific Ocean and the gulf coasts. The ridge is formed by granite that cooled from magma deep below the surface of the Earth as much as 80 million years ago. Now exposed, the granite is the same rock formation that reinforces much of the peninsula's central spine. The process of erosion active at the tip of the peninsula follows a predictable pattern in which sea caves first are enlarged by surf impacting a ridge from opposite directions. Deepening of sea caves continues until tunneling from both sides meets to form a sea arch. As erosion continues, the arch's sidewalls widen beyond the strength of the overlying roof. This stage ends with the collapse of the arch, which results in the formation of a sea stack fully separated from the rest of the ridge. Several large sea stacks marking Land's End at Cabo San Lucas show that multiple sea arches preceded the surviving arch in time. Various sea caves found farther to the west along the ridge at or near sea level indicate that future sea arches now are under excavation by the wind and waves.

All the granite exported from the sculpted sea caves, arches, and stacks, eventually is reduced to sand enriched in silica, plagioclase, and other minerals. Under the right geological conditions, the beach sand may be cemented to form sandstone. The original photograph from which Fig. 2.3 was drawn is shown in Plate 1, with the Cabo San Lucas arch and sea stacks viewed from sea looking to the south. Although it was a sunny day and the sky was clear, the wind and surf were ceaseless on both sides of the peninsula.

Diversity of Gulf Rocky Shores

Many outstanding locations reflect the diversity of modern rocky shores in the Gulf of California. Several examples are marked on the accompanying site

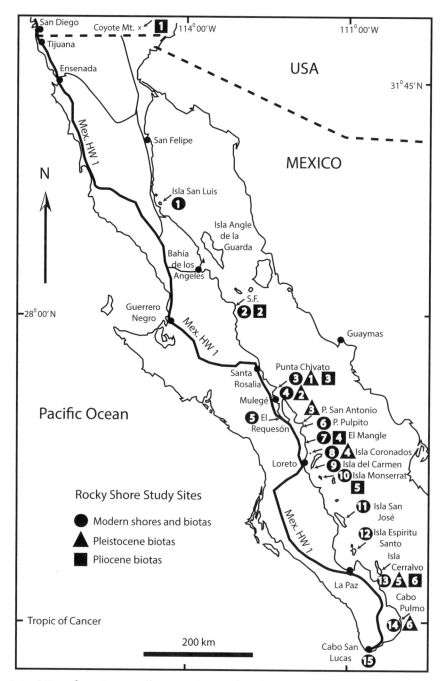

Fig. 2.4. Map showing study sites for rocky shores in the Gulf of California.

map (Fig. 2.4). Circled numbers 1 and 15 identify the previously described sites at Isla San Luis and Land's End at Cabo San Lucas. In between, there occur a succession of many picturesque places where sea cliffs and associated rocky-shore biotas may be studied. Granite sea cliffs from 80 to 120 m in height stand guard at opposite ends of a 2.5-km long sandy beach at San Francisquito (site #2).

At Punta Chivato (site #3), the sea cliffs rise tens of meters off the water and are formed by

andesite layers representing former volcanic flows now uplifted and tilted to the west. The landmark known as El Sombrerito on the harbor at Mulegé (site #4) shows a hat-like profile eroded in gabbro that represents the uplifted and exposed conduit of a former volcano. At El Rquesón (site #5), a small island at the end of a tombolo is composed entirely of andesite. Rocky-shore biotas are especially rich at this locality (see under the next heading). Punta El Púlpito (Fig. 2.5, site #6) is another well-known coastal landmark formed by rhyolite that rises

Fig. 2.5. Rhyolite igneous rocks at Punta El Púlpito.

Fig. 2.6. Limestone rocky shores at the south end of Isla del Carmen.

Fig. 2.7. Sandstone formation on the east shores of Isla San José.

abruptly to an elevation of 140 m above sea level. The jagged spires of a similar rhyolite formation are found at Punta Mencenares on the south side of the bay at San Basilio (Plate 1).

Continuing southward, the bay at El Mangle (site #7) exhibits a long rocky shore edged by Pliocene limestone (Plate 1). Sea stacks on the northeast coast of Isla Coronados show tilted layers of dark basalt (Plate 1, site #8). The south end of Isla del Carmen is formed by uplifted sea cliffs composed of Pleistocene limestone that abut against older sea cliffs eroded in andesite (Fig. 2.6, site # 9). On Isla Monserrat (site #10, Plate 1) the west coast features a long coastal shelf cut into an-

desite agglomerate. The agglomerate looks superficially like conglomerate, but it is entirely igneous in origin with pieces of volcanic rocks carried in a surface flow. One of the few sandstone formations in the Gulf of California fortifies a prominent rocky spur on the east side of Isla San José (Fig. 2.7, site #11). The sand in this Pliocene formation was eroded from a nearby former granite upland. Some of the most picturesque rocky shores occur on Isla Espíritu Santo, where narrow bays penetrate far inland along faults offset between older granite and younger andesite (site #12). Granite sea cliffs form much of the shoreline around Isla Cerralvo (site #13), although schist contributes to some parts the northwest coast as a prominent metamorphic rock.

Fig. 2.8. Sculpted granite rocks at Cabo Pulmo.

Also, layers of Pleistocene and Pliocene conglomerate are exposed at localities near the south end of this island. Some of the most accessible stretches of the peninsular shoreline occur at Cabo Pulmo (site #14), where the waves erode whimsical shapes in the granite (Fig. 2.8).

Windward and Leeward Shores

During the 1940 expedition to the Gulf of California by Steinbeck and Ricketts, extensive collections of intertidal invertebrates were made from exposed (windward) rocky shores and more sheltered (leeward) rocky shores around islands and other points along the peninsular coast. Backus et al. (2009, their Table 2.4) re-examined the distribution of marine invertebrates from 162 species collected at several localities under contrasting windward and leeward conditions. One of the most exposed collecting sites with 69 species was studied at Puerto Refugio on the northwest end of Isla Angle de la Guarda, whereas one of the most protected sites with 67 species was studied at Puerto Escondido south of Loreto. Many species are exclusive either to exposed or sheltered rocky shores. Analysis of the results from this early work is too great to fully summarize here. However, it can be noted that the chitons (polyplacophorans from the phylum Mollusca) were the most widely distributed species from among the high-energy rocky shores under review. All but one among the seven species listed by Ricketts was found to inhabit exposed rocky

shores with vigorous wave shock. In contrast, another widely distributed group of invertebrates that includes 10 species of sea stars (phylum Echinodermata), showed almost no preference for exposed over sheltered rocky shores. While the chitons are herbivores generally specialized for a rough-water habitat on rocky shores, the sea stars are the top invertebrate predators capable of moving back and forth from more exposed to more sheltered settings.

Hayes et al. (1993) performed one of the most detailed census studies on rocky-shore biotas anywhere in the Gulf of California at Isla Requesón in Bahía Concepción (Fig. 2.4, site #5). Although situated far within the bay, the north side of the islet is fully exposed to waves from the north because the bay opens in that direction. Data on the location and size of populations among 30 different species was collected using grids that allowed for accurate counts to be registered. Barnacles (*Balanus* sp.) crowd the upper intertidal zone on the north-facing shore, below which is an intermediate zone extensively encrusted by coralline red algae. Within this middle zone, the arc shell (*Arca pacifica*) is abundant (Fig. 2.9). Scattered colonies of stony finger corrals (*Porites panamensis*) and more extensive colonies of soft corals (*Palythoa* sp.) occupy the lowest intertidal zone. In contrast, a different and larger barnacle (*Tetraclita affinis*) is more common in the upper intertidal zone on the south-facing rocky shoreline, whereas a rock-encrusting

Fig. 2.9. Arc shells (*Arca pacifica*) at El Requesón.

oyster (*Saccostrea palmula*) occurs abundantly in the middle intertidal zone. In spots where the red mangrove (*Rhizophora mangle*) thrives, this oyster is found encrusted on the plant's prop roots. The cup and saucer gastropod (*Crucibulum spinosum*) also is very common in the middle intertidal zone. Present but not common, a small grazing gastropod (*Nerita funiculata*) occupies open spaces among the barnacles. In the lower intertidal zone from the south side, various green algae and a purple sponge (*Verongia aurea*) show a diffuse coverage.

At Cabo Pulmo (Fig. 2.4, site #14), the low granite shores present another location where the study of intertidal faunas is made easy by ready access. Notably, populations of a larger grazing gastropod (*Nerita scabricosta*) cluster together on outcrops where they are exposed for many hours during low tide (Fig. 2.10).

Fig. 2.10. Live grazing gastropods (*Nerita scabricosta*) at Cabo Pulmo. Coin is 2.4 cm in diameter.

Fossil Examples and Paleoecology

An important conclusion from the study by Hayes et al. (1993) is that most of the 30 species covered in the El Requesón survey possess the necessary hard skeletal parts for potential fossilization. Such a result is a critical key in the search for fossils that represent former windward and leeward habitats in the gulf's rocky-shore ecosystem.

Pleistocene relationships. Localities with fossils from Pleistocene rocky shores are marked by num-bered triangles in the accompanying map (Fig. 2.4). During the last inter-glacial epoch of the Late Pleistocene (approximately 125,000 years ago), sea level was much higher than today. This means that former sea cliffs of Late Pleistocene age are likely to be exposed on terraces now elevated and slightly more inland. The following review summarizes information from six Upper Pleistocene sites. This reflects only a small sample of the region's great potential for additional research on the paleoecology of Pleistocene rocky shores.

Based on the study by Libbey and Johnson (1997), the most northern of the six sites occurs near Punta Chivato behind Playa La Palmita (site Δ#1). Housing developments along the beach now make it difficult to find some of the spots, but good examples of former sea cliffs remain intact and undisturbed. An Upper Pleistocene sea cliff with 2 m of vertical relief exhibits as many as 15 species of rock-encrusting invertebrates. Similar to the modern setting at El Requesón, one of the common fossils cemented in place by coralline red algae is the ark shell (*Arca pacifica*). More inland at an elevation 11 m above sea level is a small, former island composed of red-colored andesite surrounded by white carbonate sediments (Fig. 2.11). Sizable with an area of 12,500 m², the feature is low in topography making it hard to distinguish except for the change in color and slight rise in elevation when passing on the road that crosses the thin paleoisland. For this reason, the authors named the site Isla Fantasma. Often preserved with articulated shells nestled in growth position among andesite cobbles, three species of bivalves appear on the north side of the paleoisland. The most abundant are the venus shells (*Chione californiensis* and *Periglypta multicostata*), although a lucinid shell (*Codakia distinguenda*) also is present in growth position on the margin of the paleoisland. In contrast, many conches (*Strombus galeatus*) are crowded near the shoreline on the paleoisland's south side. The northern side is interpreted to have been the windward side, absorbing higher wave energy.

The brim of El Sombrerito at Mulegé (site Δ#2) encircles that landmark with a terrace 12 m

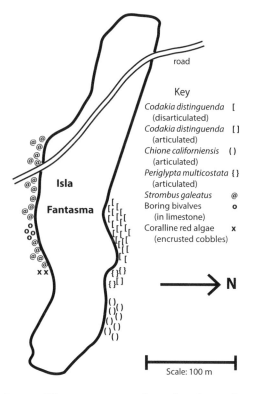

road

Key

Codakia distinguenda   [
(disarticulated)
Codakia distinguenda   []
(articulated)
Chione californiensis   ()
(articulated)
Periglypta multicostata   {}
(articulated)
Strombus galeatus   @
Boring bivalves   o
(in limestone)
Coralline red algae   x
(encrusted cobbles)

Isla

Fantasma

→ N

Scale: 100 m

Fig. 2.11. Pleistocene rocky island at Playa La Palmita south of Punta Chivato.

above sea level, and some the gravel on the terrace includes marine fossils from the last interglacial epoch of the Late Pleistocene (Ashby et al., 1987). The outer shores of the Concepción peninsula south of Mulegé typically trace the equivalent 12-m terrace continuously incised in andesite over a great distance. The same nestled bivalves found in growth position on Isla Fantasma occur among the cobbles and boulders along the prominent Concepción peninsula terrace. Farther south inland from Punta San Antonio (site Δ#3) is an abandoned embayment with a steep conglomerate cliff composed of tightly fitted granite and andesite boulders. A detailed census of the fossils preserved in growth position on this Upper Pleistocene rocky shore was conducted by Johnson et al. (1999) using the same survey techniques as Hayes et al. (1993) based on grid samples. In addition to articulated venus shells (*Chione californiensis* and *Periglypta multicostata*) and the lucinid shell (*Codakia distinguenda*) commonly wedged among the boulders, the mussel (*Modiolus capax*) was the most abundant bivalve counted. Also found at this locality, a turbinate gastropod (*Turbo fluctuosus*) and neritid gastropod (*Nerita funiculate*) are regarded as typical represen-

tatives of the upper intertidal zone. The local paleogeography indicates that the rocky embayment near Punta San Antonio was open to the north and subject to strong wave activity.

Continuing to the south, the Upper Pleistocene lagoon on Isla Coronados (site Δ#4) features an outer barrier formed by andesite islets with a narrow opening that allowed tides to flush the lagoon. Most of the time, the barrier was partly emerged above sea level but could be over-washed during storms that arrived from the south (see Chapter 5). According to a survey of the paleolagoon (Johnson et al., 2007), colonies of the stony finger coral (*Porites panamensis*) occur as fossils attached directly to the inside slope of the rock barriers. In this case, a reef structure is preserved intact, but it also represents an example of a fringing reef in a rocky-shore setting. A similar setting is found at the south end of Isla Cerralvo (site Δ#5), where a mix of finger corals and the cauliflower coral (*Pocillopora* sp.) grew as thickets with individual colonies cemented directly onto cobbles and boulders of granite, basalt, and schist (Tierney and Johnson, 2012). In this case, a diverse fauna of mollusks that includes the expected venus shell (*Periglypta multicostata*) and turbinate gastropod (*Turbo fluctuosus*) attests to the lower intertidal nature of the setting.

Another Pleistocene boulder bed is developed as an apron in front of the granite rocky shore at Cabo Pulmo (site Δ#6), where fossil oysters are preserved cemented onto the vertical granite surface (Fig. 2.12). At this site, some fossil corals oc-

Fig. 2.12. Pleistocene oysters encrusted on granite at Cabo Pulmo.

cur among the granite boulders but they are not attached to the boulders and appear to have been transported landward from a Pleistocene reef located farther offshore.

Pliocene relationships. Localities with fossils from Pliocene rocky shores are marked by numbered squares in the accompanying map (Fig. 2.4). The Pliocene Epoch lasted for two and three-quarters million years, starting about 5.3 million years ago. This interval of geologic time concluded at the start of the succeeding Pleistocene Epoch about 2.6 million years ago. The beginning of the Pliocene Epoch commonly is the time frame during which the early Gulf of California was widely flooded by seawater. The northern part of the gulf appears to have experienced a more complex history that involved widespread deposition of gypsum, most likely before the start of the Pliocene. In the following review, information from six Pliocene sites is summarized. This sample reflects only a small part of the region's rich potential for future research on the paleoecology of Pliocene rocky shores. Many embayments and coves along the entire gulf coast remain yet to be explored.

The most northern of the six sites is located at Coyote Mountain in southern California across the border with the USA (Fig. 2.4, site □#1), where the Salton Trough brought an extension of the Pliocene Gulf of California into contract with rocky shores composed of marble. Several different kinds of organic borings are preserved in the marble. These were made by variety of marine organisms that included rock-boring clams (*Lithophaga* sp.), rock-boring sponges (*Entobia* sp.), and rock-boring bristle worms (*Maeandropolydora* sp.), as well as circular pits eroded by sea urchins (Watkins, 1990). Fossil barnacles also are preserved attached to the marble surface. In some places, wave-cut notches left a signature to show the high-stand in Pliocene sea level.

A locality that includes many body fossils occurs at Punta San Francisquito (site □#2), where a great Pliocene embayment occupying 10 km² was eroded in granite (Johnson and Ledesma-Vázquez, 2009). Excavation of the large San Francisquito

basin by wave mechanics was enhanced by weaknesses in the granite caused by intersecting fracture planes oriented with maximum effect against wave trains refracted onto an east-facing shore. At the onset when marine water first flooded into the basin, oyster colonies (*Ostrea fischeri*) became affixed to the granite floor. Subsequently as sea level rose through the Middle Pliocene, other mollusks such as the bittersweet shell (*Glycymeris maculata*) became abundant among the granite cobbles and boulders that formed an apron flanking the inner east margin of the basin. Although subtidal in habitat, an extensive brachiopod population (*Laqueus erythraeus*) dominated the south ramp below the basin rim. The spines of fossil sea urchins are abundant along the north side of the basin, suggesting that echinoderms took advantage of a more sheltered setting.

Probably the best-studied Pliocene rocky shores on the Gulf of California are located among a cluster of paleoislands at Punta Chivato (site □#3). Rising seas at the start of the Pliocene gradually drowned as many as five separate islands centered around tilted andesite blocks (Johnson and Ledesma-Vázquez, 2001; Johnson 2002). Evidence for the concept of marine onlap is demonstrated in the east-facing cliffs of Punta Chivato, where conglomerate and limestone formations are juxtaposed against a former rocky shore composed of solid andesite (Fig. 2.13). Part of the outcrop is hidden by talus that accumulated as rock falls below the cliff face, however the greater part of the exposure exhibits andesite with a southern slope overlain by conglomerate composed of andesite cobbles and boulders eroded from the adjacent parent rock.

Through its onshore track, the conglomerate shows as many as six reversals that project into the adjacent limestone to the south. The six excursions of conglomerate are labeled R1 to R6, signifying a half dozen, short-lived pauses in rising sea level or perhaps even brief falls in sea level. The letter R stands for a **marine regression**, which in this case represents a shift in the conglomerate from an overall onshore track to shorter intervals of offshore retreat. The horizontal and vertical scales

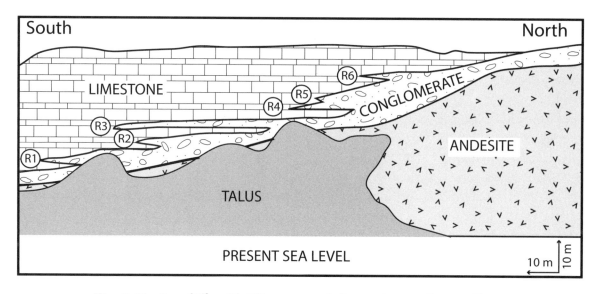

Fig. 2.13. Sea cliffs with Pliocene rock formations at Punta Chivato.

affixed to the diagram (Fig. 2.13) confirm that main phase of marine onlap stretched over a lateral distance of 200 m and represents a minimum 50-m rise in relative sea level. Within the limestone, most of the fossils are represented pecten shells. Due to limited access at the cliff face, it is hard to assess the fossil content within the conglomerate. However, at other places on the north side of the Punta Chivato promontory the most common fossils within the basal conglomerate include the same bivalve (*Glycymeris maculata*) as found in the rocky-shore conglomerate in the San Francisquito basin. Another arc shell (*Barbatia reeveana*) also is plentiful. The living representatives of this clam employ organic threads (byssus) to adhere to a rock surface. Yet another common fossil found in the Punta Chivato conglomerate is a heavily armored sea urchin (*Clypeaster bowersi*) that was capable of withstanding strong wave action.

Limestone that follows after the basal conglomerate on the north side of the Punta Chivato promontory typically includes a stony coral (*Solenastrea fairbanksi*) that first appeared in the Gulf of California during earliest Pliocene times (Foster, 1979). The geologic range of this fossil confirms that the underlying Punta Chivato conglomerate can be correlated with the Zanclean Stage of the Lower Pliocene. Whereas the overlying limestone on the east side of the Punta Chivato promontory (Fig. 2.13) is dominated by fossil pectens, the co-

eval limestone on the south flank of the promontory contains abundant fossil oysters. Variations in the surrounding limestone reflect different invertebrates that lived somewhat offshore the rocky islands and not the organisms that lived in immediate contact with the rocky coast. However, those different limestone layers strongly reflect the paleoecology of opposing windward and leeward shores in an Early Pliocene archipelago.

Another outstanding locality where Pliocene rocky shores have been studied in much detail is located south of Punta El Mangle (Fig. 2.4, site □#4). The wide limestone shelf filling part of a sheltered bay abuts against a former andesite rocky shore on the east flank of a shield volcano called the Cerro Mencenares volcanic complex that covers an area of 150 km[2] on the Gulf of California. The locality is geologically significant for several reasons (Johnson et al., 2003). The first is that the sedimentary rocks near El Mangle can be dated to 3.3 million years in age on the basis of a volcanic tuff layer inter-bedded with the succession. This means that the sedimentary formations can be correlated with the middle part of the Piacenzian Stage belonging to the Upper Pliocene. Secondly, the paired limestone and conglomerate beds at El Mangle can be shown to trace a phase of marine onlap followed by a regressive phase of marine offlap (Fig. 12.14). In particular, the onlap can be traced for a lateral distance of 310 m across a shelf eroded by wave action

18

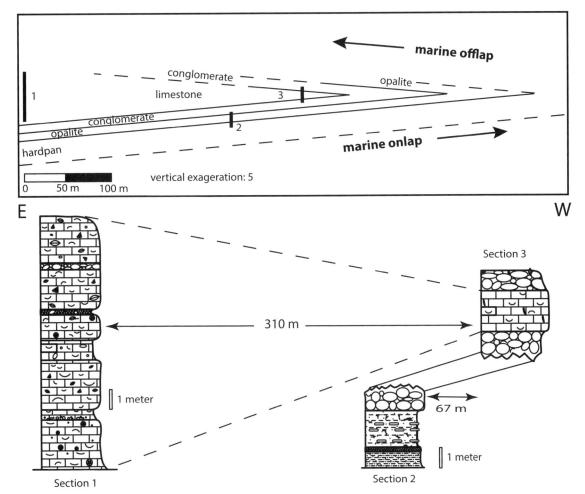

Fig. 2.14. Schematic showing marine onlap and offlap in Pliocene strata at El Mangle.

along the andesite paleoshore. It is rare to see such a good example of a transgressive-regressive cycle captured so precisely by a layer of conglomerate. The conglomerate, itself, contains cobble-size clasts of andesite that are heavily encrusted by coralline red algae and by clusters of barnacles. The spines from fossil sea urchins are a common component in pockets of limestone dispersed among the cobbles. Other unusual aspects of this succession are covered in Chapters 7 and 10.

A third interesting aspect of the study site is shown by a fault that offsets the Piacenzian sequence to the north at El Mangle. El Coloradito Fault is the name formally given to this feature (Johnson et al., 2003). The fault trace appears to be blocked by the Mencenares volcanic complex, but in the opposite direction it runs offshore parallel to active transform faults in the center of the Gulf of California. Because the fault post-dates the primary sedimentary sequence, it suggests that lateral

movements in the network of gulf faults were initiated in this region sometime after 3.3 million years ago.

Isla Monserrat is an intermediate-size gulf island that is difficult to reach, but features one of the most interesting examples of a Pliocene rocky shore (Fig. 2.4, site □#5) on account of the insight it gives in regard to tectonics. The study site is located in the south-central part of the island, where a wedge of Upper Pliocene limestone from the Piacenzian Stage meets a paleoshore marked by andesite that can be traced for a distance of 750 m (Johnson and Ledesma-Vázquez. 2009; Johnson, 2014). Fossils of the larger grazing gastropod (*Nerita scabricosta*) are abundant along the geological contact and they attest to the intertidal nature of the thin limestone layer at this site. Conglomerate beds exposed elsewhere include the same fossil bittersweet shell (*Glycymeris maculata*) found abundantly on Pliocene rocky shores at San Francisquito (site □#2) and Punta Chivato (site □#3). From a

19

tectonic point of view, it is curious that the highest exposure of Isla Monserrat limestone now sits at an elevation 204 m above sea level. Red andesite rocks that denote the crown of the paleoisland rise in height for another 40 m nearby. The setting makes a powerful statement on the dramatic vertical uplift of Isla Monserrat during the last 3 million years when only a small part of the tilted fault block that defines the island barely was above sea level.

More accessible from the peninsular mainland of Baja California than Isla Monserrat, Isla Cerralvo is a larger island with another Pliocene rocky shore superbly exposed at Los Carillos (site ☐#6) near the southeast end. Contact between granite and Upper Pliocene conglomerate traces a former shoreline for a distance extending about 100 m (Johnson et al., 2012). Capped by a thin layer of limey sandstone, the sedimentary sequence at this site forms a well-defined ramp deposit that dips naturally 7° into the sea just as it once did during Pliocene time. The larger grazing gastropod (*Nerita scabricosta*) is the most abundant fossil preserved within pockets of silica sand in the basal conglomerate (Fig. 2.15). Fossil barnacles are rare, but occur attached to granite cobbles. At the south end of the outcrop, smooth granite mounds (up to 10 m across) protrude through the sandstone exposed

Fig. 2.15. Pliocene gastropods (*Nerita scabricosta*) from Los Carillos on Isla Cerralvo.

like the skerries they once were when washed by rising seas 3 million years ago. Several basalt dikes cut through the granite, and were already exhumed above the granite surface as vertical walls that acted as natural groins to hold a mix of granite and basalt cobbles in place against erosion by Pliocene waves and longshore currents. Beautiful on its own account as a former rocky shore in the process of being reclaimed by the sea, Los Carillos presents an entirely realistic window on the Pliocene past.

Chapter 3

Coral Beds and Reefs

Introduction

Among the ecosystems spread around planet Earth, perhaps that of coral reefs enjoys the highest profile in our human consciousness. Coral reefs are popular destinations for visiting ecotourists due to their beauty and the diversity of other life forms that take refuge within. In this sense, the stony corals that build a ridged reef framework are regarded as a classic **umbrella species**. In other words, the reef structure formed by only a few key species provides the support and shelter for a host of other species otherwise unlikely to thrive in that particular place. Another reason why coral reefs often come to our attention is the scale of environmental degradation that affects the ecosystem's fragile health. The entire system comes under threat of destruction when key players no longer are able to cope with higher water temperatures due to global warming or increased acidity due to increasing amounts of carbon dioxide absorbed into the water column.

Living coral reefs are hard to study because the system's physical architecture tends to be massive and covers over the relationships that originally were in operation when the reef first began to develop. It is like viewing a great building from the outside without access to details hidden within the basement foundation. In this regard, the study of fossil coral beds and reefs is important due to the fact that contemporary erosion of rock formations now on dry land provides a cross-section through the inner parts of the coral superstructure. Yet another advantage to the study of fossil coral beds and reefs is that the distribution of those ecosystems thousands and even millions of years ago can be compared with what we find in place today. When it can be shown that distributions were more widespread in the past, it means that the geographic limitations under which coral beds and reefs now exist have changed from earlier times. The penin-sular lands and islands on the Gulf of California are exciting places to explore, because fossil coral beds and reefs are abundant, widespread, and well preserved.

Stony Corals and Soft-bodied Relatives

Reef-building corals together with sea fans and other soft-bodied relatives belong to the phylum Cnidaria. Paleontologists often refer to the same group as members of the phylum Coelenterata. The biological name for this phylum comes from specialized stinging cells (**cnidocytes**) concentrated in tentacles that distinguish these animals from all other invertebrates. The name used by paleontologists (**coelenterata** = hollow gut) relates to another trait particular to these animals, which is a blind gut with no opening other than the mouth. The stony corals and their soft-bodied relatives are stationary carnivores anchored to the seafloor. They are aggressive feeders that capture prey through the use of tentacles armed with stinging cells.

The form and function of a basic coral polyp may be modeled on a soft-bodied relative: the sea anemone (Fig. 3.1). This animal has a stout

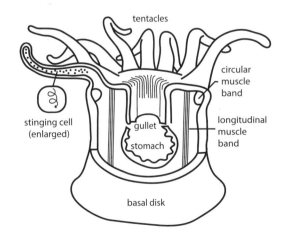

Fig. 3.1. Cross-section through the polyp of a sea anemone.

trunk attached to the seafloor by strong muscles in the basal disk. Upwards, the tubular body expands into an oral disk with multiple tentacles that surround the mouth. Tipped with a sharp harpoon-like projection, the stinging cells individually are coiled like a rope beneath the surface membrane of the tentacles. The cells are fired at once a foreign body brushes against that particular spot. Contact with the prey results in the injection of a toxin. The gullet in a sea anemone is lined with specialized cells (flagella) that convey the stunned prey to the stomach. When digestion is complete, the remains of a meal must be regurgitated back through the mouth.

Relatively large anemones of the kind that inhabit tidal pools around the Gulf of California are able to ingest prey, such as small shrimps. Most are solitary animals, but some also form colonial mats. Few stony corals in the Gulf of California are represented by solitary individuals, as for example the mushroom coral (*Cycloseris elegans*). Reef-forming corals are exclusively colonial and the individual polyp is quite small. It means that the prey consumed by reef-forming corals is correspondingly small, mainly tiny animals called **zooplankton** that drift about in water currents. Polyps in colonial corals sit within individual skeletal cups (**corallites**) spread over the surface of the entire stony colony (**corallum**). During the daytime, tentacles belonging to individual polyps are withdrawn below the surface of the corallite. Feeding occurs during the nighttime, and the extruded tentacles appear like a delicate hairy cover on the surface of the colony. Many reef-building corals supplement food intake through symbiosis with single-celled algae (zooxanthellae) that live within the tissue of the coral polyps. The algal cells produce oxygen and food through photosynthesis that is shared with the polyp and the polyp provides algal cells with carbon dioxide and other needed materials.

Limitations on Coral Distribution

Reef-building corals are regulated in their global distribution by sensitivity to conditions in the physical environment. Among the most important are water temperature and water clarity. The optimum water temperature for healthy coral growth is between 23° and 25° Celsius (73° to 77° Fahrenheit). American geologist James D. Dana (1813-1895), who traveled widely among the islands in the Pacific Ocean during the United States Exploring Expedition (1838-1842), was the first to clearly understand the importance of water temperature in the healthy development of coral reefs. His studies indicated that a minimum temperature of 20° C (68° F) in surface ocean water was critical to the establishment and maintenance corals in a reef setting. He acknowledged that individual coral colonies outside a reef setting might survive in waters as cold as 18° C (64°F), but that coral reproduction was much diminished.

Water clarity is equally important to reef corals, because sunlight is required by the symbiotic algae living within coral tissue to carry out the process of photosynthesis. Fine sediment dispersed in the water column cuts down on available sunlight, but also has the potential to foul the feeding tentacles of coral polyps. In addition, the symbiotic algae associated with corals are unable to withstand water temperatures that long remain above 25° C. Corals that lose their algae become bleached of color and they are less likely to survive. Last but not least, corals are marine creatures unable to tolerate fresh water. It means that coral reefs will not thrive in places where streams or rivers enter the ocean from land, even if the water temperature and water clarity are otherwise suitable.

The temperature of surface seawater decreases on a latitudinal gradient from the Earth's equator to the poles as moderated by atmospheric circulation. Coral reefs are tropical to subtropical in nature, typically occupying territory between the latitudes of 30° N and 30° S. However, they are poorly represented on the western margins of continents, where oceanic gyres bring cooler waters towards the equator. This relationship is shown for part of the Pacific Ocean off the coast of North America, where the California Current sweeps from north to south and causes deviations in the temperature gradient as shown for a typical win-

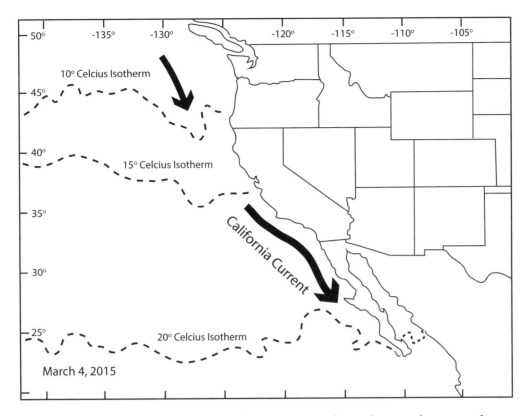

Fig. 3.2. North Pacific Ocean off North America with isotherms shown in degrees Celcius (oblique view).

ter day in March (Fig. 3.2). Wiggles in the lines representing the 10°, 20°, and 30° isotherms are deflected southward closer to the coast, most notably along the outer shores of the Baja California peninsula. Sea anemones are a common inhabitant of tidal pools from Canada's Vancouver Island to the tip of the Baja California peninsular, but stony corals are prohibited from colonizing those colder outer shores. In more sheltered waters belonging to the southern Gulf of California, however, the 20° isotherm follows a pronounced northward excursion. Stony corals do manage to survive as separate entities far to the north within the Gulf of California, but the reef-building corals are restricted collectively to a zone south of the 20° isotherm.

Life Expectancy of a Reef System

The life expectancy of reef corals may span several decades. Growth bands typically preserved within a given coral colony record the age in years, where the bands reflect subtle seasonal changes. Growth is mostly directed upward but also outward, as individual colonies reach into shallower water. Up-

ward growth halts with arrival at the intertidal zone, because coral species are sensitive to periods of subaerial exposure and especially to higher levels of ultraviolet radiation in the sunlight. When no space remains for upward growth, coral colonies expand laterally and become more densely crowded. A coral reef may continue to grow upward generation after generation over thousands of years, where the seafloor on which it is seated experiences long-term subsidence. Otherwise, reef structures and the corals that build them are prone to destruction by major storms and hurricanes.

Different outcomes in the history of a coral reef over time can be illustrated by plotting the effects of exposure to the elements on one axis against the effects of transport (or removal) from the site of original growth on the other (Fig. 3.3). The scenario for model 1 (lower, left corner) represents a situation in which a coral colony sustains continuous growth unaffected by exposure and is not impacted by forces that otherwise might interrupt growth in any way. Such a reef structure may be buried in place without any damage and become preserved

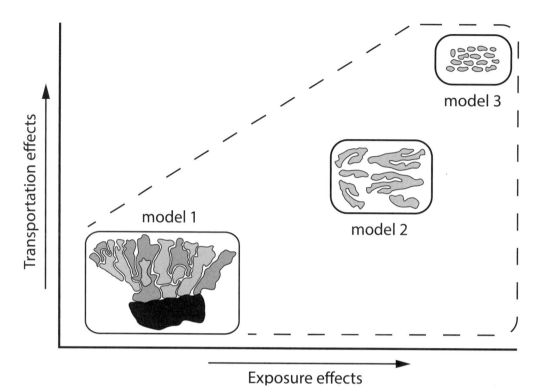

Fig. 3.3. Growth and decay of reef corals under the effects of erosion with time.

intact as a fossil reef. Model 2 (center right in the plot) denotes a case in which branches of a coral colony are broken off and removed by waves or currents to another locality. Under favorable circumstances, growth may resume with the branches seeding a new reef structure. More likely, however, the individual polyps living on those branches will quickly expire, but the debris may be preserved as a post-mortem deposit in the fossil record.

Model 3 (upper right corner) reflects long-term, post-mortem exposure to the elements and ongoing transportation of material away from the original site of reef growth. In this case, much of the structure is reduced to rubble that may be carried farther offshore by strong currents as coral gravel or washed onshore to a beach setting. Likewise, this scenario is liable to preservation in the rock record, although the end result no longer reflects a living reef but perhaps an offshore sand bar inhabited by entirely different organisms. Overall, the three models contribute to a better understanding of reef growth in more exposed (windward) settings as distinct from more sheltered (leeward) settings.

Living Reef Systems in the Gulf of California

The accompanying map for the Gulf of California (Fig. 3.4) shows the limited range of localities (circled numbers) where living coral reefs occur today. As summarized by Reyes-Bonilla and López-Pérez (2009), many years of research have been devoted to the reefs in Bahía San Gabriel at Isla Espíritu Santo (site #1) and Cabo Pulmo (site #2) in the southern gulf region. These are regarded as true reefs that exhibit a complex structure supporting a diverse community of organisms. Skeletal frameworks are dominated by a single kind of cauliflower coral (*Pocillopora* sp.) capable of building structures up to 2 m in height. Because the average growth rate of this coral in the lower Gulf of California is 3 cm/year (Reyes-Bonilla and Calderón-Aguilera, 1999), it may be estimated that the reef at Cabo Pulmo represents no less than 70 years of continuous development. At least 10 other coral species are sheltered within the Bahía San Gabriel and Cabo Pulmo reefs together with a rich assemblage of many crustaceans, mollusks, echinoderms, and fish species.

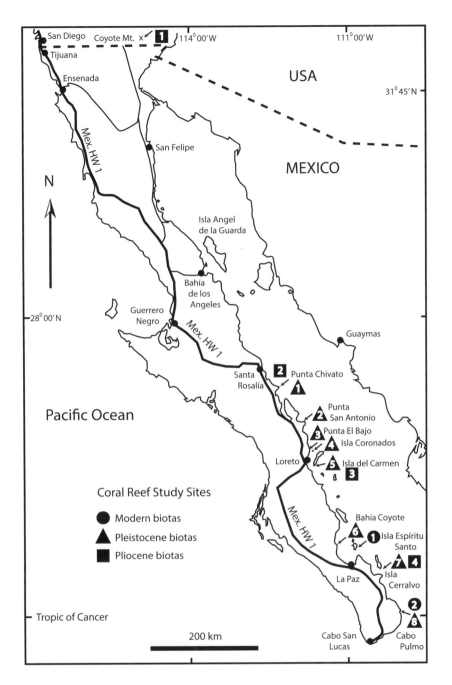

Fig. 3.4. Map showing study sites for coral beds and reefs in the Gulf of California.

At El Paso between Isla Espíritu Santo and Isla Partida, storm waves deposited an immense berm against the sea cliffs on the southeast side of the smaller island (Fig. 3.5A). The massive berm exceeds one meter in height and extends along the shore over a distance of tens of meters. This feature includes scattered andesite cobbles and a sparse mixture of shells, but otherwise is dominated by broken and eroded branches of the cauliflower coral (*Pocillopora* sp.) as shown by the photo inset (Fig. 3.5B). The example corresponds to model 2 in the preceding discussion on the effects of exposure and transportation in the natural history and life expectancy of a coral reef (Fig. 3.3).

Other kinds of corals survive in places farther to the north, but occur as loose aggregates in coral beds that bear no resemblance to reef structures. For example, the mushroom coral (*Fungia* sp.) inhabits shallow waters off Isla Monserrat and the cup coral (*Phyllangia consagensis*) tracks as far north as Bahía los Angeles. Both of these are stony

Fig. 3.5. Coastal berm on Isla Partida formed by reef debris (A); Inset (B) showing branches from the cauliflower coral (*Pocillopora* sp.).

corals, but represent individual (non-colonial) forms. The places where these and other corals occur in non-reef settings are too numerous to indicate on the study map.

Fossil Examples and Paleoecology

Changes in sea level over the past 5 million years are the main reason why extensive coral reefs and other coral beds preserved in limestone layers are readily available for study on land around the Gulf of California today. Tectonic uplift of the former seabed is another factor that contributes to this situation. Examples of Pleistocene and Pliocene reefs or coral beds are marked on the map by numbered triangles and numbered squares, respectively (Fig. 3.4).

Pleistocene relationships. It is noteworthy that fossil reefs from the Late Pleistocene occur two and a half degrees of latitude farther north than where the corresponding living reefs are found today in Bahía San Gabriel off Isla Espíritu Santo. This is equal to a north-south separation of about 280 km. Fossil Pleistocene reefs range from Punta Chivato

in the north to Cabo Pulmo in the south (DeDiego-Forbes et al., 2004; Johnson et al., 2007; López-Pérez 2012).

The fossil reef at Punta Chivato (Fig. 3.4, site Δ#1) is sheltered on the south side of a protective ridge. Finger corals (*Porites panamensis*) stand 15 cm high, three generations of which are cut by an arroyo through the reef's center. The outer part of the reef rests on gravel with imbricated cobbles and pebbles, whereas the inner part sits directly on Pliocene limestone. This locality features the most diverse fossil fauna presently known from the Gulf of California with 50 shelly species preserved in growth position among the coral branches. Bivalves are articulated, with a venus shell (*Chione californiensis*) and lucinid shell (*Divalinga eburnean*) among the most abundant. Less common is an arc shell (*Barbatia reeveana*) and a turbinate gastropod (*Turbo fluctuosus*). A census based on multiple samples yielded an average diversity of 12 species per grid using a quadrat 25 x 25 cm moved laterally through the reef exposure. Laboratory analysis for radiometric isotopes dates the reef to the last interglacial epoch about 118,000 years ago.

Unusual for the east-facing orientation, a former coastal inlet preserves a long rocky shoreline at Bahía San Antonio near San Nicolás. A cove approximately 750 m² in area occurs at end of the inlet (site Δ#2), where a single generation of the finger coral (*P. panamensis*) achieved a colony height of 18 cm. Here, the dominant colony shape is more massive and less branched in form as at Punta Chivato. A fauna of 18 marine invertebrates is associated with the reef. Among the most abundant is the turbinate gastropod (*T. fluctuosus*) and a venus shell (*Chione tumens*). Using a quadrat 25 cm on a side for each sample, multiple grid samples yielded an average diversity of 12 species per grid. A radiometric date for this site is not available, but the age conforms to the last interglacial epoch based on physical correlation with adjoining marine terraces (Johnson and Ledesma-Vázquez, 1999).

Across from Isla Coronados on the peninsular mainland, Punta El Bajo (site Δ#3) features a north-facing shoreline with a 50-m exposure through a Late Pleistocene reef. The reef is formed by the finger coral (*P. panamensis*), although the robust morphology adopted by the colonies is unique to the Gulf of California. Individual branches up to 10 cm long have a slender attachment with the core that broadens into a wide frond at the tip. A single generation of corals achieved an average colony height of 45 cm and equal girth. Some corals are attached to andesite boulders, but most sit directly on a limestone surface. Colonies from the middle of the reef are upright in growth position, but others are tilted 30° to the sides. The reef's biodiversity is low with only 8 shelly species. The most abundant is the turbinate gastropod (*T. fluctuosus*) and a venus shell (*Chione californiensis*). The corals at this locality have not been dated using radiometric isotopes, but nearby marine terraces close to the same elevation are dated between 80,000 and 105,000 years in age.

The largest and best preserved fossil reef from the Gulf of California covers an area of 10,600 m² and resides at an elevation close to 12 m above sea level at Cañada Coronados on the south side of Isla Coronados (Fig. 3.4, site Δ#4). Within a setting interpreted as a semi-enclosed lagoon at the foot of a Pleistocene volcano, the reef structure is estimated to include 13,000 coral colonies preserved in growth position (Johnson et al., 2007). The corals (exclusively *P. panamensis*), grew attached to andesite cobbles and small boulders that paved the floor of the lagoon (Fig. 3.6). Originally, the

Fig. 3.6. Colonies of Pleistocene corals (*Porites panamensis*) in growth position on Isla Coronados.

lagoon had a sandy bottom filled with the debris of crushed rhodoliths (see Chapter 5), but a major rain storm was responsible for a wash-out event that brought an influx of rocky material into the lagoon off the nearby slope of the volcano. Corals need a firm place for anchorage in order to grow, and a close examination of the pavement surface shows that corals rapidly took advantage of the opportunity to colonize the lagoon, as represented by an example of an immature colony preserved in place (Fig. 3.7). Essentially, the lagoon pavement served as rocky platform that attracted other sorts of organisms besides the corals. As illustrated (Fig. 3.8), coralline red algae encrusted some of the andesite pebbles, as did a common mussel (*Modiolus capax*), and various gastropods characteristic of a shallow-water setting (left: *Turbo fluctuosus*, below: *Nerita bernhardi*, and right: *Acanthina tuberculata*) also populated the new surface. These and many other marine invertebrates continued to live within the branches of the maturing corals, which grew to a maximum height extending one meter. Laboratory analysis for radiometric isotopes dates the reef

Fig. 3.7. Detail of immature *Porites* colony attached to andesite cobble.

Fig. 3.8. Pleistocene fossils (gastropods, bivalve, and coralline red algae) from the reef site on Isla Coronados.

to the last interglacial epoch between 127,000 and 121,000 years ago. No more than a single generation of coral growth is preserved on site, and apparently the reef expired when colonies reached the top of the water column with nowhere else to expand.

Branching corals (*P. panamensis*) are preserved in growth position at sea level and inland 11 m above sea level on the south shore of Isla del Carmen (site Δ#5). Colonies exposed on a bedding plane at the coast are up to 60 cm in diameter. They occur as a single generation directly attached to a limestone surface. At least two generations of colony growth are exposed in cross section at the 11-m elevation. Growth occurred on a substrate of loose coral fragments, but some colonies up to

45 cm in height are attached to disarticulated shells of a spiny oyster (*Spondylus princeps*). Life within this reef was sparse, limited to marine worms that left dwelling tubes 10-15 cm in length cemented to coral branches. No radiometric age determination is available for this locality, but the 11-m elevation is consistent with marine terraces throughout the region correlated with the last inter-glacial epoch.

Bahía Coyote is located 70 km north of La Paz (site Δ#6), where a Pleistocene embayment sheltered extensive reefs formed by closely branched corals (*P. panamensis*). Dense coral build-ups occur at more than a half-dozen spots over a distance of 15 km (DeDiego-Forbes et al., 2004). Fossil corals preserved in growth position sit on coarse bioclastic sand, oyster debris, and sparse cobbles. Colonies 25 cm to 50 cm in height developed massive accumulations with multiple generations that attained a maximum thickness of 8 m. Infiltration by shelly invertebrates seeking shelter appears to have been minimal in these massive *Porites* beds.

The most intensely studied Pleistocene reefs from the Gulf of California are found at the southwest corner of Isla Cerralvo (Fig. 3.4, site Δ#7), where a close succession of five reefs followed after each other in a vertical sequence (Tierney and Johnson, 2012). The colonies in each reef structure are anchored to a pavement of cobbles and small boulders stabilized by extensive growth of coralline red algae (Fig. 3.9). Most of the colonies were established by the finger coral (*P. panamensis*), but the cauliflower coral (*Pocillopora* sp.) also is represented. The growth form of the dominant coral is

Fig. 3.9. Colonies of Pleistocene corals (*Porites panamensis*) from Isla Cerralvo.

uncharacteristically dense, lacking distinct branches, and the average height is only about 25 cm. Each cycle of reef growth is terminated by a fresh influx of cobbles that reflect a surprising range of rock types from granite to basalt to schist. Smooth and rounded, these rocks are interpreted as having been washed from the island's interior through ephemeral streambeds during high-intensity rain events that left small fan deltas on the shore.

Rain events of the magnitude required to transport a large volume of river cobbles on Isla Cerralvo probably resulted from the passage of hurricanes. Longshore currents associated with strong seasonal winds consistently removed some cobbles from the fan deltas between major storms and carried them southward along the coast to reside in places where the nearby rocky shores consist only of granite. The stack of five reefs that follow one another in succession have a combined thickness of 2.5 m, which indicates that part of the Cerralvo coastline experienced an equal amount of subsidence measurable on account of the fact that reef growth never exceeds mean sea level. Superb exposure of the reef cycles both in vertical cross section and in planar view allowed for a detailed census of the other marine invertebrates living within (Tierney and Johnson, 2012). Details include a record of barnacles (*Hexacreusia durhami*) and specialized bivalves (*Lithophaga* sp.) that commonly bored into the coral colonies (Fig. 3.10). Laboratory analysis for radiometric isotopes dates a sample from one of the lower reefs to the last interglacial epoch 122,000 years ago. Subsequent to development of the final reef cycle, the entire sequence was uplifted to assume its present exposure on the coast.

At Cabo Pulmo (Fig. 3.4, site Δ#7), several species of the dominant reef-building coral (*Pocillopora damicornis, P. meandrina, P. capitata,* and *P. lobata*) occur as Pleistocene fossils (López-Pérez, 2012). However, they are not preserved in place as part of a surviving reef structure. A largely intact coral colony belonging to *Pocillopora* can be seen wedged into a crevice in the coastal granite at Cabo Pulmo (Fig. 3.11). The colony was broken off the reef structure in which it formerly grew and was

Fig. 3.10. Detail showing ingrown barnacles and cavity left by a bivalve (arrows).

transported onto the rocky shore at Cabo Pulmo. As such without greater damage to the corallum, the example suggests a post-mortem scenario somewhere between models 1 and 2 (Fig. 3.3). Close by this spot, fossil coral debris more closely resembling the collection of corals in the modern berm on Isla Partida (Fig. 3.5) occurs trapped within recesses of the granite.

Fig. 3.11. Colony of a Pleistocene cauliflower coral (*Pocillopora sp.*) on the rocky shore at Punta Pulmo.

29

Pliocene relationships. The distribution of corals is sporadic in Pliocene strata that date from 5 million years ago, or even farther back in geologic time. However, the fossil record demonstrates that corals entered the Gulf of California and migrated as far north as southern California where corals unrelated to reef structures are found in the Coyote Mountains (Fig. 3.4, site □#1). According to López-Pérez (2012), fossils from the Imperial Formation of southern California include several genera of corals (*Porites, Siderastrea. Diploria*, and *Solanastrea*). They all reached that area from the south at a time when sea level stood appreciably higher than today. In particular, the species *Solanastrea fairbanksi* is regarded as having originated in the Mediterranean area during the Miocene and migrated first to a Caribbean, and from there through the straits of Panama before the connection to the Pacific Ocean was closed by the Isthmus of Panama about 3.5 million years ago (Foster, 1979).

Solanastrea fairbanksi also occurs widely in coral beds on the north side of the Punta Chivato promontory (site □#2), where it lived exclusively on the windward side of rocky Pliocene islands (Johnson and Ledesma-Vázquez, 2001). The coral has a hemispherical shape from 15 to 20 cm in diameter (Fig. 3.12) and is typically associated with conglomerate beds that accumulated in the San Marcos Formation as rocky-shore deposits. The rounded shape of the coral was able to withstand the impact of strong waves, but it never formed reef

Fig. 3.12. Rounded Pliocene coral (*Solanastrea fairbanksi*) from the north shore of Punta Chivato.

structures. This fossil also is reported together with other corals (*Siderastrea* and *Favia*) in similar conglomeratic deposits at Puerto de la Launcha (site □#3), on the northern, windward coast of Isla del Carmen (López-Pérez, 2012).

The final Pliocene locality considered in this review comes from the west coast of Isla Cerralvo at Paredones Blancos (Fig. 3.4, site □#4), where a true reef structure formed by one of the cauliflower corals (*Pocillopora* sp.) takes on the characteristic branching shape of a large candelabra (Fig. 3.13). The corals are preserved in a limestone layer that follows a succession of conglomerate layers and limestone dated by index fossils as typical for the middle Pliocene (Emhoff et al., 2012). Preservation showing details of corallites from this reef coral is very good (Fig. 3.15), but research still waits to be done on the reef and its associated biota at this locality.

In combination, what is known about the distribution of Pleistocene and Pliocene corals in the Gulf of California provides a fascinating glimpse into the past. All marine invertebrates residing in the gulf today come from immigrant stock, because the body of water mostly did not exist prior to six-million years ago , or as much as 12 million years ago in the far south. That some of the first coral immigrants arrived from Europe and the Caribbean tells us much about the potential for species to move from place to place through larval transport by the prevailing ocean currents of that time. The story is yet more intriguing, given that the most common kind of the finger coral in the gulf (*Porites panamensis*) is considered to have evolved from ancestors that immigrated from places in the western Pacific (López-Pérez, 2012), long after the Isthmus of Panama chocked off immigration from the Caribbean. Much remains to be learned.

Fig. 3.13. Branching Pliocene coral (*Pocillopora* sp.) from Paredones Blancos on Isla Cerralvo.

Fig. 3.14. Detail showing the filling of corallites (negative mold) on the surface of a cauliflower coral (*Pocillopora* sp.) from Paredones Blancos. Individual corallites are 0.4 cm in diameter.

Chapter 4

Clam Flats

Introduction

Shellfish known as clams in the common vernacular are more properly referred to as bivalves in the scientific literature, in particular as representatives belonging to the class **Bivalvia** within the phylum Mollusca. As such, the more formal name refers to an exoskeleton with two outer shells (valves) held together by a hinge and strong internal muscles. The related name "mussels," given shellfish from temperate waters, reflects the importance of this group as a food source for humans. In addition to bivalves, other common mollusks include the chitons, gastropods (sea snails), and cephalopods (octopuses, squids, and relatives). Overwhelmingly, these are marine creatures, although freshwater clams and air-breathing snails (slugs) made successful transitions from the sea to terrestrial environments. Except for cephalopods, the mollusks that thrive in marine settings today mostly remain armored by protective shells composed of calcium carbonate ($CaCO_3$). For paleontologists, the distinction is important because the shelled mollusks left a robust mark on the fossil record.

Worldwide, marine gastropods belong to the most diverse class, estimated to include over 60,000 living species. Next come the marine bivalves, with a little more than 8,000 living species, followed by the chitons with a little less than 1,000 living species. The cephalopods are the least diverse, represented by about 800 living species worldwide. Within the Gulf of California, the number of known mollusk species follows the same order, but with notably fewer numbers. The gulf's marine gastropods number 1,534 species, the bivalves 566 species, the chitons 57 species, and the cephalopods 20 species (Brusca and Hendrickx, 2010). These numbers reveal only a preliminary accounting, because marine biologists have yet to conduct a more thorough census of the entire re-

gion. For example compared with the number of species living today in a well-studied region such as the Mediterranean Sea, the actual number of bivalve species living in the Gulf of California is estimated to be almost 1.5 times the number of known species (Brusca and Hendrickx, 2010).

Simple species lists fail to take into consideration rank abundance. Bivalves rank high due to the productivity of only a few species responsible for enormous populations that have spawned, lived, and died in the Gulf of California over the past several millions of years. This high level of productivity is recognized independently in two ways. One sign is shown by kitchen middens left by the earliest human inhabitants along gulf shores, but also by great shell piles left by small-scale commercial operators working, for example, around Bahía Concepción. The second and far more impressive sign is purely geological in nature. Extensive limestone formations composed almost entirely of bivalves that date back to the Pliocene are found many places around the Gulf of California. These deposits are nothing short of monumental in scale and they reflect conditions that developed long before humans had any impact on the region's ecology.

Bivalve Form and Function

All clams possess an exoskeleton that consists of two valves, but subtle variations in the shape of those shells reflect a great difference in life style of different bivalve species. Much may be learned about the habits of clams by examining the inside surfaces of their shells. Radically different living patterns are demonstrated by two kinds of bivalves commonly found in the Gulf of California (Fig. 4.1).

Typical venus shell. The first example is the prolific chocolate shell (*Megapitaria squalida*), much appreciated throughout Baja California for its culinary value. As a typical venus shell (Family Veneridae), the adult has an oblong shape with a maximum length of 12 cm and height of 9.5. The shell designated as the right valve (Fig. 4.1A), is a mirror image of the opposite (left) valve. Several features are visible as scars or imprints on the inner surface of a clean valve. Two muscle scars of equal size form shallow pits towards the ends of the shell, showing the attachment place of the anterior and posterior adductor muscles. These are strong contractile muscles that close the shell and keep it tightly closed under duress. When the muscles are relaxed, the shell springs open through the response of the resilium, located in a groove below the umbo. Composed of material with properties much like vulcanized rubber, this feature forms a wedge between the two shells near the hinge line and exerts a naturally expansive force.

The pallial line is another significant feature, which is marked by a linear trace etched into the shell between the anterior and posterior muscle scars. It reveals the terminal growth of tissues that adhere directly to the shell. With time, the location of this trace shifts outward as the clam grows. A distinct indentation in the pallial line appears near the posterior muscle scar (Fig. 4.1A). This notch marks the position of a siphon that functions like a breathing tube to convey oxygenated water (and food items carried by the water) from the outside powered by the internal pump-action of the animal's gills. The siphon is the chocolate shell's lifeline to the outside for an animal that lives beneath the sand hidden from view except for the opening at the surface where the siphon pokes out. Marine biologists regard this clam and its relations as **infaunal suspension feeders.** The term has a dual purpose, telling that the clam's habitat is subsurface in nature and that sustenance is obtained by sieving small food particles (phytoplankton) suspended in the water.

The outer surfaces of the chocolate shell's perfectly matched valves are smooth and without ornamentation. Active digging achieved by a muscular foot keeps to keep the animal hidden in the

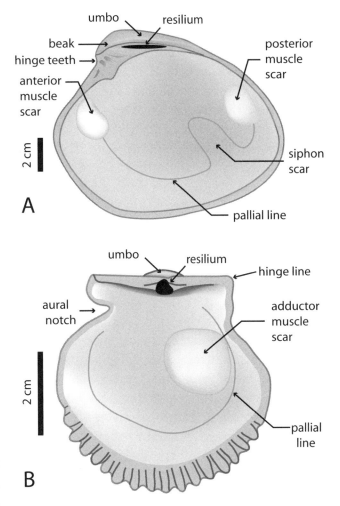

Fig. 4.1. Valves of common clams showing internal views. A) Chocolate shell (*Megapitaria squalida*). B) Small scallop (*Argopecten circularis*).

sand below the surface. The exterior smoothness of the shell offers no resistance as the animal slips through the sand. Shallow sand flats close to shore are the preferred habitat of the chocolate shell, but it may inhabit sediments up to 160 m below the water surface.

Typical pecten shell. Another prolific bivalve from the Gulf of California is the pecten (*Argopecten circularis*), commonly referred to as a scallop and also much appreciated for its culinary value. As implied by the scientific name, the shell has a circular shape, but is framed by a straight hinge line and a set of ridges and furrows that leave a scalloped pattern at the shell's distal edge. Like the chocolate shell, the inner surface of a clean pecten shell reveals a pallial line marking the outer extremity of tissues. In contrast to the chocolate shell, the interior of the pecten shell reveals a single, large adductor muscle scar (Fig. 4.1B). Furthermore, there is no sign of an indentation on the pallial line marking the location of a siphon. The pecten has no need for a siphon, because it does not live in the sand below the water-sediment interface. Instead, it spends much of the time resting on the surface of the sand. The pecten also has a resilium located near the hinge line and it functions exactly the same way as it does for the chocolate shell. However, a subtle but significant difference in the pecten is that its two shells are not a perfect mirror image of one another. One valve is flatter in profile than the other. Its opposite is distinctly more curved in profile.

Whereas the chocolate shell has an advantage in hiding from potential predators, the pecten is exposed at the surface to hunters like the starfish or octopus. However, the pecten has a keen sense of smell and is fully aware of a predator's approach. Forewarned of danger, the pecten flexes its powerful adductor muscle, thereby snapping the shell closed. Swift closure of the shell forces a sudden expulsion of water through the anterior, scalloped margin of the shell. This action sets off an equally strong counter-action that propels the pecten into the water column. The term "jet propulsion" is appropriate to describe this kind of flight. Moreover, like the cross-section on an airplane wing with an upper surface that is slightly arched in profile, the difference in flow over the curved pecten valve in contrast to the opposite flatter valve creates lift just as it does in an airplane. Essentially, pectens are capable of short flights through the water column when obliged to flee from potential predators.

Marine biologists regard the pecten as an **epifaunal suspension feeder.** The term refers to the pecten's habit of living exposed on the surface of the sea floor, but also points to the invertebrate's means of sustenance feeding on minute phytoplankton suspended in the water column. Many pectens achieve a larger size, but *Argopecten circularis* has a maximum adult size of about 5 cm in diameter. The species has a range that extends from the lower tidal zone to offshore depths of 135 m.

Limits to Clam Distribution

Few studies are available on the population density and physical limits of clam distribution in the Gulf of California. Water depth and associated factors such as wave energy, water temperature, and food resources are regarded as the main controls. Some show a preference for more vigorous wave energy near the intertidal zone, whereas others seek calm conditions in deeper, offshore waters. Thus, the size of suitable habitat area appears to be the major condition that determines where and how many bivalves may prosper in any given area.

As noted above the chocolate shell (*Megapitaria squalida*) does well in relatively shallow sand flats, but may inhabit finer sediments in waters to a depth of 160 m. In a study regarding the chocolate shell in Bahía Juncalito south of Nopoló (Villalejo-Fuerte et al., (1999), a systematic search for the bivalve was conducted using a gridded network of sample stations along transects that extended 500 m offshore to a maximum depth of 12 m. It was found that on average, the population density of live chocolate shells is about 1.5 animals per square meter. Related to this statistic is the fact that the chocolate shell has a normal life span of five to six years, as may be determined by counting major growth lines recorded on the valves but also through more sophisticated tests looking at seasonal variations in oxygen isotopes absorbed within the valves (Skudder et al., 2006). Shell samples from the Loreto area were tested for seasonal variations in temperature that indicate changes on the order of nearly 10°C (18°F).

As discussed, the common pecten (*Argopecten circularis*) thrives atop sand flats but also may inhabit deeper zones to a water depth of 135 m. Scallops can be harvested in the Gulf of California by commercial trawlers, but little information is available on catch size related to specific areas at specific depths of water. Historically, some information on the harvesting of scallops is reported from places such as Isla Angel de la Guarda (MacKintosh, 2008), where teams of helmet divers operated during the early 1970s to manually collect the species *Pecten vogdesi* for commercial purposes. Divers were limited by how deep they could safely operate, but in a relatively short period of time virtually all scallops to a depth of 18 m (about 60 feet) were harvested from the eastern shelf off Angel de la Guarda.

A crude method to check on areas where clams are most abundant is to walk the beaches, taking a census of the kinds of shells washed ashore by the tides and occasional storms. By consulting bathymetric charts that map variations in water depth off specific beaches, it is possible to estimate the size of habitat area occupied by the populations of those species. Although water depth rapidly falls off between the peninsular mainland and the gulf's central axis, the size of habitable sand flats above the 50 m isobath off places like the north shores of Punta Chivato, Isla del Carmen, and Isla Monserrat is considerable. Based on a live count for as few as one or two bivalves for a given species per square meter, the corresponding population can be estimated as between 10,000 to 20,000 bivalves within an area as small as 10 km². What happens to all these clams as they grow to maturity and expire under natural conditions is another question worthy of attention.

Clam Populations in Life and Death

A range of life and post-mortem scenarios designed to account for an infaunal clam like the chocolate shell (*Megapitaria squalida*) is explored schematically in Figure 4.2. Model 1 shows individuals as part of a functional population safely concealed below the surface of a sand flat. Surface waves intensified by the seasonal north winds, for example, exert a greater effect on those clams living in shallow water close to shore at the edge of the tidal zone. Vigorous wave and surf activity have the power to shift sand and uncover clams. A healthy adult clam, however, will resist exposure by digging back into the sand. For that part of the population located in deeper water farther offshore, the surface waves from winter winds are less invasive as a force capable of shifting a large volume of sand. Those individuals need not work so hard to maintain a safe hiding place. In any case, model 1 presumes that a healthy clam population remains unaffected by such a disturbance. Hence, model 1 is positioned near the graph's point of origin with no ill effects of exposure or transportation.

Model 2 assumes a central place in the graph (Fig. 4.2), where chocolate clams are represented

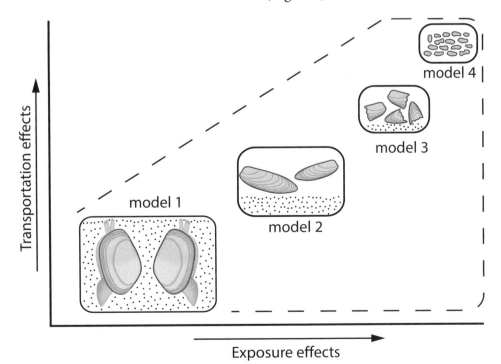

Fig.4. 2. Exposure and transportation effects on infaunal clams over time.

after death as shells lying empty on a sandy surface. The valves from a single individual remain whole, but are disarticulated from one another. Clams that are not preyed upon by other animals like skates or harvested by humans will reach maturity and arrive at a natural death, at which time the muscles that hold their shells together undergo decay. Winter waves or major storms that uncover dead clams will leave many shells strewn across the surface of the seafloor. Some few may be transported by waves all the way to a beach to arrive in pristine condition.

Models 3 and 4 occupy positions on the graph (Fig. 4.2) influenced by progressively severe effects of exposure and transportation. The longer disarticulated valves sit exposed on the bottom, the greater the incidence of water turbulence during which shells collide against one another and suffer breakage. Once broken into pieces, the parts undergo further wear as they rub against each other and are polished by friction with surrounding sand. Ultimately, a large volume of shell bits and carbonate sand derived from mollusks may accumulate on north-facing beaches (model 4), where the material remains trapped in place except for rip tides that transport some beach sand back offshore. The same winter winds that agitate strong surf on north-facing beaches also impact beaches by blowing a fine fraction of beach sand inland where it settles in coastal dunes. Especially in gulf regions from Isla Monserrat northward, the beaches and coastal dunes are dominated by organic material derived from clams and other mollusks (see Chapter 6).

Clam Flats in the Gulf of California

Identified by circled numbers on the map in Figure 4.3, the showing of clam flats in the Gulf of California is by no means comprehensive. A sample of eight outstanding areas provides a brief introduction to spots where hunting for clams at low tide and for beach combing may be practiced with success. At the head of the gulf between San Felipe and the Colorado River delta (site #1), long coastal bars called **cheniers** are almost entirely dominated by the disarticulated valves of a single clam with the scientific name *Mulinia coloradoensis* (Avila-Serra-

no et al., 2009). It is a small bivalve with an adult shell length of 5 cm, and the species is exclusive to the delta region where high levels of nutrients were formerly introduced by the river prior to the conservation of river water by a system of upstream dams. Today, populations of *M. coloradoensis* are in severe decline, but trillions of shells belonging to this estuarine bivalve remain in circulation subject to long shore transport and deposition along the shores directly south of the delta.

Located 7 km south of the village at Bahía de los Angeles (site #2), a 4-km beach with an east-west orientation marks the termination of the partially enclosed embayment. The locality is a natural trap for shells washed onto the beach from the southern half of Bahía de los Angeles by wind-driven waves sweeping sand flats approximately 25 km² in area. Berms behind the beach are piled with shells dominated foremost by venus shells (*Protothaca grata* and *Chione californiensis*), but also the little heart shell (*Cardita affinis*). Oyster shells and some gastropods also are present, but rare. In particular, the dominant venus and little heart shells are known to occupy intertidal sand flats, but also range into more subtidal waters.

Another source area for shells transported in bulk volume to beaches is found north of Punta Chivato (site #3) within the Ensenada el Muerto. Beaches west of Punta Chivato and near El Rincón south of Isla San Marcos are plastered with worn bits of shell derived mostly from bivalves living in the adjacent sand flats. Subtidal sand flats within the 50-m isobath in this area easily amount to 50 km². South of Mulegé, the beaches around Cerro El Gallo (site #4) are the natural repository for shells sorted by waves from subtidal sand flats that cover as much as 75 km² in area (Skudder et al., 2006).

Within Bahía Concepción to the south, the tombolo at El Requesón and connected islet (site # 5) represent an ideal place to explore contrasting environments where bivalves thrive. Among the clam species that populate the intertidal sand flats astride the tombolo, the pen shell (*Pinna rugosa*) is especially abundant in the island's leeward shel-

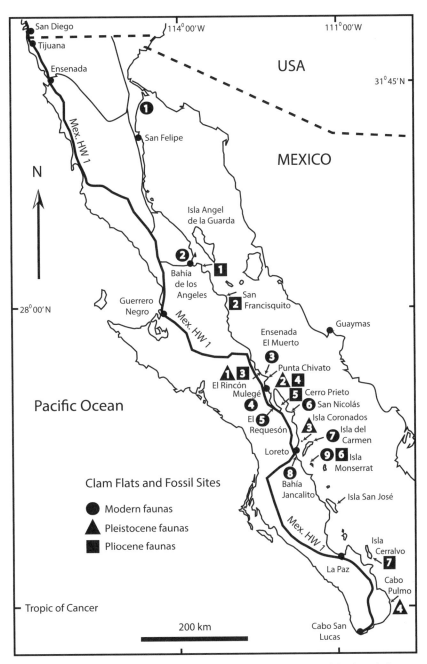

Fig. 4.3. Present-day and fossil clam flats in the Gulf of California.

ter. This fan-shaped bivalve lives oriented vertically in the sand with only a small part of the shell visible above the surface (Fig. 4.4). It is regarded as a **semi-infaunal suspension feeder**. The bivalve is exclusively intertidal in habit and judging from the mounds of empty shells behind the nearby mangroves, it has been extensively harvested at this locality. Somewhat like the scallop, the pen shell has a single adductor muscle that functions to close the shell. The muscle is small (< 2 cm in diameter) compared to a typical scallop, but the pen shell is commonly utilized as a food source in Baja Cali-

fornia. On the exposed, windward side of Isla Requesón, large piles of empty arc shells (*Arca pacifica*) also attest to the harvest of this species (Fig. 4.5).

Beaches with an east-west orientation south of San Nicolás (site #6) are favored with deposits of the chocolate shell (*M. squalida*). In comparison to the beaches near Mulegé, however, the adjacent source area of subtidal sand flats is estimated as only about 10 km². North-facing pocket beaches on Isla del Carmen (site #7) also are positioned to

Fig. 4.4. Pen shell (*Pinna rugosa*) hidden in growth position at El Requesón.

Fig. 4.5. Shell pile with abundant arc shells (*Arca pacifica*) at El Requesón.

trap abundant shell debris from the adjacent sand flats above the 50-m isobath approximately 50 km² in area. Likewise, an area of 90 km² that surrounds the islets of Las Galeras lies within the 50-m isobath north of Isla Monserrat (site #8). The north beach on Isla Monserrat is oriented in such a way to capture abundant shell debris, typically dominated by the bittersweet shell (*Glycymeris maculata*) and chocolate shell (*M. squalida*).

Fossil Examples and Paleoecology

As in the case of fossil corals discussed in Chapter 3, changes in sea level in unison with patterns of tectonic uplift during the past 5 million years are the chief agencies behind the accessibility of bivalve deposits located on land today. Examples of Pleistocene and Pliocene clam beds are marked on the map by numbered triangles and numbered squares, respectively (Fig. 4.3).

Pleistocene relationships. A deposit of loose shells that dates from the last inter-glacial epoch about 125,000 years ago (Johnson et al., 2009), blankets a patch of ground roughly 250 m² close to El Rincón (Fig. 4.3, site Δ#1). The shell deposit is cut by a dirt road to reveal an exposed thickness of 0.5 m (Fig. 4.6). Present but uncommon, a turbinate gastropod (*Turbo fluctuosus*) suggests that the deposit has an intertidal to shallow-water origin. Fossil clams fully dominate the deposit, only a few of which remain with valves articulated. By implication, much

of the material was transported some distance from where the invertebrates once lived. Clams with an infaunal habit are represented, the most abundant being the chocolate shell (*M. squalida*) supplemented in lesser numbers by venus shells (*Chione californiensis* and *C. tumens*) as well as the jackknife shell (*Tagelus affinis*).

Similar shell beds that date from the last interglacial epoch are more widely dispersed behind Playa La Palmita (site Δ#2) on the leeward side of the Punta Chivato promontory. A series of small Pleistocene oyster colonies occupy ridges eroded in Pliocene basement rocks at this locality (Johnson et al., 2009). The colonies are preserved intact with

Fig. 4.6. Pleistocene deposit near El Rincón with abundant chocolate shells (*Megapitaria squalida*).

several generations of shells fully articulated. An exhaustive survey of the other shelly fauna at this locality reveals one of the highest diversity assemblages ever documented from the Pleistocene of Baja California, including more than 95 species of bivalves (Cintra-Buenrostro et al., 2002). As at El Rincón, infaunal species represented by the venus and jack-knife shells are commonly represented in this assemblage.

A mixed assemblage of infaunal and epifaunal bivalves occurs in the limestone escarpment outside the Pleistocene coral lagoon on Isla Coronados (site Δ#3). The assemblage features 17 species of bivalves, many of which remain intact as articulated individuals (Johnson, 2014, p. 150). The most abundant clams include a venus shell (*M. squalida*), a rock oyster (*Spondylus crassiquama*), a pecten (*Lyropecten subnodosus*), and an arc shell (*Anadara multicostata*). Other common species include additional venus shells (*Chione californiensis* and *Periglypta multicostata*) together with a lucinid shell (*Codakia distinguenda*). A fauna of 20 gastropods is intermixed with the bivalve fauna, but the sea snails are much less abundant than the bivalves. Overall, the Pleistocene mollusks at this locality are densely concentrated but well preserved, suggesting that the assemblage was not far removed from where its various members formerly lived. Within the Pleistocene lagoon on Isla Coronados, coarse carbonate sand derived from crushed rhodoliths filled the space prior to development of the coral reef. Bivalves such as the lucinid shell (*Codakia distinguenda*) commonly found refuge in this setting. Many such shells are preserved in growth position (Fig. 4.7), now exposed in layers below the reef.

Cited previously in Chapter 2 on rocky shores, the locality at Cabo Pulmo (Fig. 4.3, Δ#4) features Pleistocene oysters preserved in growth position attached to granite. Other kinds of fossil bivalves are not conspicuous at this locality.

Pliocene relationships. Several outcrops of Pliocene limestone bearing abundant bivalves are reviewed in this section. The most northern occurs 12 km southeast of Bahía de los Angeles at Ensenada El

Fig. 4.7. Pleistocene lucinid shell (*Codakia distinguenda*) in growth position.

Quemado (Fig. 4.3, site □#1). During the middle Pliocene, the headland at Punta Quemado was separated from the peninsular mainland by a 250-m wide marine passage that extends laterally for a distance of 2 km (Johnson et al., 2009). Dacite conglomerate sits on the same igneous basement rocks and partially fills the channel. Above, follows a 5-m thick limestone that incorporates quartz granules eroded from the adjacent dacite rocks. Fossil oysters (*Ostrea megadon*) occur near the opening of the channel onto Ensenada El Quemado. Farther within, a large bittersweet shell (*Glycymeris gigantea*) is the dominant bivalve with a density of about three articulated individuals per square meter as measured vertically against the outcrop (Fig. 4.8). Today, this epifaunal species typically inhabits the seafloor at a water depth between 7 and 13 m. By implication, the paleochannel was flooded to an equivalent depth, but the introduction of sediment was rapid enough to bury the large bittersweet shells after minimum post-mortem exposure.

The great Pliocene embayment at San Francisquito (Fig. 4.3, site □#2) previously treated in Chapter 2, is notable for two kinds of fossil clams. The exposed floor of the bay was colonization initially by small oysters (*Ostrea fischeri*) that took advantage of elevated spots on the granite surface. As the bay continued to fill with seawater, profuse

Fig. 4.8. Pliocene outcrop with large bittersweet shells (*Glycymeris gigantea*) at Ensenada El Quemado.

populations of a small bittersweet shell (*G. maculata*) nestled among granite cobbles adjacent to the granite coastline.

Located roughly midway between Santa Rosalía and Mulegé (site □#3), a vast table land of Pliocene limestone covers an area no less than 9 km² around El Rincón. It is the most extraordinary place in all Baja California where stupendous numbers of fossil clams occur. Three kinds of bivalves are super-abundant in the limestone layers that comprise a sequence nearly 50 m in thickness (Johnson et al., 2009). The by now familiar chocolate shell (*M. squalida*), is one of them, although preserved only as impressions after the original shells were lost to dissolution. Still, the evidence is impressive based on bedding planes that reveal about 40 disarticulated valves per square meter (Fig. 4.9). The scenario is consistent with model 2 (Fig. 4.2), wherein wave energy is responsible for uncovering and moving the separated valves of infaunal species across the sea floor.

The limestone cap rock at El Rincón is composed entirely of fossil oysters (*Ostrea cumingiana*). In places, the oysters form distinct mounds that rise 2 m in height (Fig. 4.10). They are tightly packed (Fig. 4.11) and often preserved as articulated individuals with a count of about 48 recumbent oysters per square meter (Johnson et al., 2009). Oysters

Fig. 4.9. Pliocene surface with chocolate shells (*Megapitaria squalida*) near El Rincón (tool for scale = 70 cm).

grow naturally cemented together in dense concentrations that conform in principle to the scenario in model 1 (Fig. 4.2), but with regard to epifaunal species. In the vicinity of El Rincón, a single square kilometer of oyster cap rock represents roughly 48 million oysters congregated on a former Pliocene seabed now elevated 15 m or more above sea level. The silty limestone below the oyster cap rock

Fig. 4.10. Mound of Pliocene oysters near El Rincón (tool for scale = 70 cm).

Fig. 4.11. Detail showing Pliocene oysters near El Rincón (pocket knife for scale = 9 cm).

contains abundant fossil pectens (Fig. 4.12, all with disarticulated valves most of which are broken into small pieces. The scenario fits model 3 (Fig. 4.2), indicative of considerable exposure and transport.

Eastward, the nearby headland at Punta Chivato (site ☐#4) represents a cluster of former Pliocene islands with andesite cores surrounded by conglomerate and limestone deposits in which bivalves are abundant. From Chapter 2, the profile given in Figure 2.13 portrays a high cliff on the east face of the promontory that entails a 35-m thick sequence of limestone layers filled with fossil pectens. The deposit rests adjacent to and on top of conglomerate eroded from an andesite rocky shore. A more accessible spot where the equivalent limestone and associated conglomerate may be explored is the oddly shaped Punta Cacarizo, also known locally as "Hammer Head Point." Unfortunately, the fossil pectens are preserved only as imprints that are difficult to identify to genus and species level. However, the general nature of the deposit conforms to model 2 (Fig. 4.2) in terms of post-mortem exposure and transport.

Around the landmark known as Cerro Prieto near the base of the Concepción peninsular (Fig. 4.3, site ☐#5), narrow limestone-filled valleys cross the igneous landscape from Bahía Concepción to almost reach the far side of the peninsula. Pliocene oysters (*Ostrea californica osunai*) are surprisingly abundant in what amounts to a maze of internal lagoons isolated from Bahía Concepción in middle

Fig. 4.12. Pliocene pectens near El Rincón (pocket knife for scale = 9 cm).

Pliocene time. Some of the fossil oysters are very large (Fig. 14.13), with a maximum length of 44 cm. Although abundant and often articulated, the oysters are not cemented to the igneous andesite at the lagoon floor, nor are they cemented together in a solid mass. Most likely, the environment within

Fig. 4.13. Large Pliocene oyster (*Ostrea californica osunai*) near Cerro Prieto.

the remote lagoons held more brackish or more hypersaline water tolerated only by the oysters. Had a through-going connection been established with the open gulf on the far side of the Concepción peninsula, marine circulation surely would have improved. As it happened, the Pliocene peninsula came close to becoming a large Pliocene island.

Silty limestone layers rich in Pliocene pecten shells (*Patinopecten bakeri diazi*) are exposed along the southwest side of Isla Monserrat (Fig. 4.3, site □#6). Some valves reach 20 cm in diameter and are well preserved (Fig. 4.14). Often they are disarticulated and fragmented. An unusual feature found at this locality is the nucleation of pecten shells packed together into spherical forms up to 70 cm across attributed to armored mud balls (Ledesma-Vázquez et al., 2007). The principal outcrop in which the mud balls occur reclines at a 6°angle that takes the form of a long slide with a change in elevation amounting to 84 m from top to bottom. The slide is regarded as a natural feature preserved intact from the time that the layers were deposited against the island's side. With pecten debris held in place by clay, the mud balls are interpreted as having formed during earthquakes that shook the island and triggered a kind of shelly, underwater avalanche that surged down slope.

While not as distinctive as the pecten deposits from Isla Monserrat, another example of a silty limestone with fossil pectens occurs as part of a ramp deposit located at Los Carillos on the southeast shore of Isla Cerralvo (Fig. 4.3, site □#7). The concentration of disarticulated pecten shells is sparse, but follows after a basal conglomerate associated with the adjacent granite shore. Unlike the unusual mud balls from Isla Monserrat that signify an offshore transfer of material, the Pliocene pec-

Fig. 4.14. Large Pliocene pecten (*Patinopecten bakeri diazi*) from Isla Monserrat.

tens at Los Carillos are interpreted to have washed shoreward during a general rise in sea level. The assortment of shells is consistent with model 3 in the generation of bivalve deposits (Fig. 4.2).

Many past and present gastropod shells from the Gulf of California are both delicate and beautiful in their architecture. Some are rare and sought after by shell collectors. By comparison many gulf clams are large and clunky. Where the bivalves may be lacking in aesthetic appeal, they surpass other mollusk relations as prolific builders of limestone formations during Pleistocene and Pliocene times. The limestone tablelands around El Rincón attest to the broad extent of former clam flats. Already in Pliocene times, the Gulf of California was an uncommonly deep body of water, much as it is now. Even so, there remain today large areas of sand flats above the 50-m isobath where bivalves proliferate both as epifaunal and infaunal dwellers. Oysters, scallops, pen shells, and especially the chocolate clams also appeal to the human palate.

Chapter 5

Rhodolith Banks

Introduction

Sunlight is required for photosynthesis by plants, whether on land or in the water. Marine algae include green, red, and brown types commonly with soft foliage anchored to the sea floor by rootlets. In principle, habitat space is subdivided according to the depth of seawater at which different wavelengths of sunlight are filtered out. The greens tend to grow more luxuriantly than the reds in shallow water under the strongest illumination. Reds outperform the greens and browns in slightly deeper water, but brown algae thrive at water depths under poor illumination where the reds and greens are less viable. Certain kinds of green and red algae are specialized for production of skeletal parts through secretion of calcium carbonate ($CaCO_3$) at the cellular level. There are no examples of brown algae capable of mineralization. Due to mineralization in some green and red algae beginning millions of years ago, a good fossil record is available for study.

This chapter is devoted to red algae that are heavily mineralized, but also exhibit growth forms with no means of attachment to the sea floor. The division **Rhodophyta** (from Greek roots for "rose-colored" and "plant") is the formal name for all red algae, whether capable of mineralization or not. Calcified algae within this group vaguely resemble corals, and hence are referred to as **coralline red algae** even though they share no affinities with marine animals. Those coralline red algae with a spherical or semi-spherical shape are called **rhodoliths** (from Greek roots for "rose-colored" and "rocks"). They roll about on the seabed under the influence of surface waves or bottom currents, thus allowing every part of the surface at least some exposure to sunlight. Shared with some other marine organisms, this kind of mobility is called **circumrotary movement**.

Rhodoliths live in vast numbers in what sometimes are referred to as **maerl beds**. Especially in places like the Mediterranean Sea, the Irish Sea, and even coastal Norway, maerl beds have a long history of study. The range in latitude of living rhodoliths is immense. Maerl beds are described from places as distant as New Zealand far south of the equator and Spitsbergen far north of the equator (Foster, 2001). Ambient water temperature is not a factor in the healthy production of most rhodoliths, but access to sunlight is critical. Seasonality with intervals of less sunlight plays a greater role in high-latitude settings than in tropical settings. Also, the angle at which sunlight strikes the surface of the Earth is more at a tangent as it enters water over latitudes increasingly distant from the Equator. The rate of growth shown by rhodoliths and their longevity may vary, accordingly. This is a topic of much interest, but data collected in a consistent way on a global basis have been slow to materialize (Foster, 2001).

Environmental Limitations

Coralline red algae respond to subtle variations in the physical environment regulated by factors indirectly associated with distance from shore and the depth of the water column above the seabed (Adey and Macintyre, 1973). More directly, such factors include the gradual decrease in water energy reaching the seabed from the surface and the gradual diminution of sunlight (Fig. 5.1). Coralline red algae encrust ledges exposed inter-tidally at the coast, as well as rocks that project through fine sediment at a greater distance offshore in deeper water. Expenditure of wave energy is greatest at the coastline, where rocky shores are subject to erosion that generates pebbles, cobbles, and even boulders. The near constant movement of cobbles and pebbles in rough water off rocky shores pro-

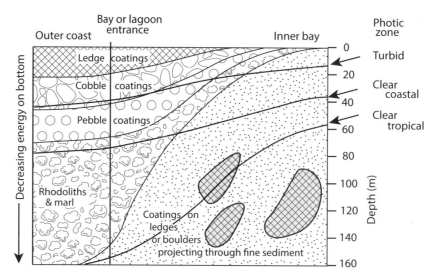

Fig. 5.1.   Sunlight and water-depth constraints on coralline red algae (after Adey and Macintyre, 1973).

vides a ready source for the nucleation of coralline red algae that accrete evenly to mimic the shape of host clasts. Larger cobbles remain closer to shore, whereas smaller pebbles may be carried farther off-shore into deeper water. Thus, the preservation of fossil rhodoliths around a pebble or cobble relates basic information about wave intensity and proximity of a former shoreline.

Rhodoliths also nucleate around small bits of shell, coral, and even pieces of broken rhodoliths. Those with no rock core may originate in an environment farther offshore in deeper water where wave energy from the surface is depleted but bottom currents are active. In a tropical setting under calm-water conditions, the water clarity is such that the **photic zone** reaches downward to a depth of 60 m or more (Fig. 2.1). On tropical continental shelves with abundant rhodoliths, such as the Abrolhos Shelf off Brazil in the southwest Atlantic Ocean (Amado-Filho, 2012), rhodolith banks commonly exceed 70 m in water depth to reach depths close to 150 m.

Shape and Surface Variations in Rhodoliths

Where rolling action by a given rhodolith is random, the result over time is equal exposure to sunlight on all surfaces. In that case, growth is evenly concentric from the center of the rhodolith outward. Where movement is constrained in two directions, a rhodolith takes on a spindle shape from back-and-forth movements. Disk-like shapes showing more growth on one side than the other tend to occur in deeper water settings where movement is influenced less by water energy and more by disruptions from marine animals living among the rhodoliths. In closer detail, the surface features of rhodoliths show variations described as **lumpy** (coarsely radiating branches with bulbous tips), **fruticose** (slender radiating branches with more open space between neighboring branches), and **foliose** (delicate fronds with a bladed appearance).

Size differences are notable, with the larger lumpy rhodoliths attaining diameters of 9 cm or more (Fig. 5.2A). Smaller rhodoliths with diameters from 4 cm to 5 cm are common (Fig. 5.2B), often showing a densely branched, fruticose growth form. In the Gulf of California, wave-agitated beds inhabited mostly by lumpy and fruticose rhodoliths develop from 3 m to 12 m below the surface (Steller and Foster, 1995; Riosmena-Rodríguez et al., 2010). Delicate fruticose and foliose rhodoliths are unable to withstand vigorous wave action unless restricted to a more sheltered setting. The example of a fossil rhodolith encrusted around a granite pebble (Fig. 5C), shows the intensity of wave action in very shallow water adjacent to a rocky shore from which the pebble was eroded. Deep-water rhodoliths in the

44

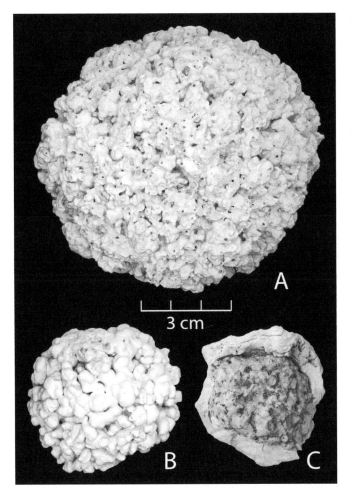

Fig. 5.2. Modern and fossil rhodoliths. A) Large rhodolith (lumpy growth) from Pichilingue. B) Small rhodolith (lumpy growth) from Punta Chivato. C) Pliocene rhodolith with granite-pebble core from San Francisquito.

Gulf of California reside on current-agitated beds up to 40 m below the surface (Riosmena-Rodríguez et al., 2010). Water energy imparted by surface waves and by bottom currents keeps rhodoliths in motion to a greater or lesser degree of frequency. Water energy also keeps rhodoliths from being buried by fine sediment that would otherwise block sunlight.

Rhodoliths and Biodiversity

According to Norris (2014), rhodoliths in the Gulf of California include six species under the genus *Lithophyllum*: (*L. margaritae, L. diguetii, L. imitans, L. hancockii, L. proboscideum,* and *L. pallescens*) among species in other genera that form mainly as crusts. Survey work summarized by Riosmena-Rodriguez et al. (2010) identifies three species besides *L. margaritae* as well represented by rhodoliths in the Gulf of California. They include *Neogoniolithon trichotomum, Mesophyllum engelhartii,* and *Lithothamnion muelleri.* A given rhodolith may enlarge through continuous growth by a single species of coralline red algae over time. It is just as likely that one species is replaced by another species that resumes growth over the surface of the same rhodolith. Annual banding is recorded by the growth of branches that radiate outward within a rhodolith. In *L. muelleri,* the growth rate has been measured as 0.6 mm/year and it is estimated that a large rhodolith from this species with a diameter of 15 cm represents a life span of 120 years (Riosmena-Rodríguez et al., 2010).

Beyond the number of coralline red algae species crowded together as rhodoliths on a marine bank, many invertebrate species live both imprisoned within individual rhodoliths and more freely among them. In this sense, a rhodolith bank is a self-contained ecosystem similar to a coral reef in support of a much wider array of marine life. Hence, rhodoliths also constitute umbrella species that provide shelter to other species that otherwise would not occupy that particular space. In a study that examined biodiversity from within individual rhodoliths (Sewell et al., 2007), it was found that 18 invertebrate species with durable hard parts capable of fossilization were self-immured. Samples mechanically broken to expose the internal contents came from modern rhodoliths stranded in the supra-tidal zone at Punta Bajo. In addition to immature clams from two species (*Barbatia illota* and *Modiolus capax*), the assemblage also included a chiton (*Stenoplax conspicua*), as well as small corals and echinoids. Other studies using a much larger sample base have identified more than 100 such **cryptofaunal** species dominated by crustaceans living among internal rhodolith branches (Riosmena-Rodríguez et al., 2010).

Among macro-invertebrates that live freely among rhodoliths, there are at least four species of crustaceans, three gastropods, and five clams

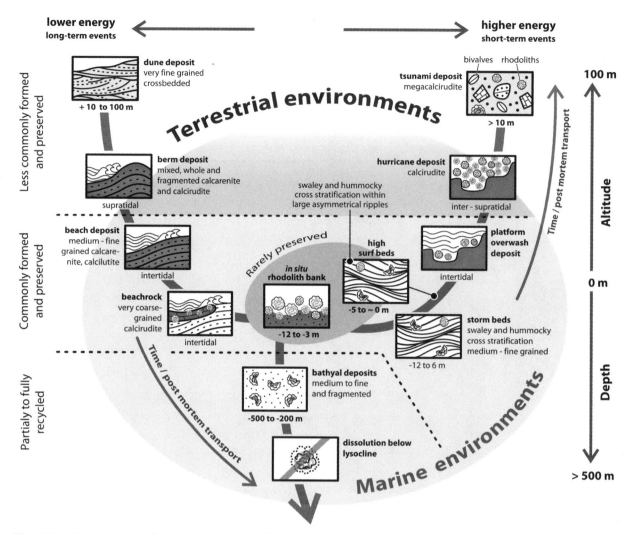

Fig. 5.3. Exposure and transportation effects on rhodoliths over time (after Johnson et al. 2012).

that currently are exploited by commercial fisheries in the Gulf of California (Riosmena-Rodriguez et al., 2010). These include scallops (*Argopecten circularis* and *Lyropecten subnodosus*) that contribute to especially lucrative markets. Unfortunately, hookah divers harvesting the scallops have done damage to the rhodolith banks as documented in Bahía Concepción. This serves notice that stronger conservation measures are needed to protect a vital natural resource in support of sustainable fisheries.

Rhodoliths in Life and Death

Due to mobility as semi-spherical objects not tethered to the seafloor like other marine algae, rhodoliths and associated organisms are subject to outcomes in life and death related to normal, day-to-day events as well as more sporadic events associated with short-term bursts of energy at higher

than usual intensities. The inherent difficulty of keeping rhodoliths corralled together on shallow banks over indefinite periods of time is central to this picture (Fig. 5.3). Ultimately, many rhodoliths are doomed to shift either landward or seaward from a habitat normally accustomed to low or moderate wave activity felt from 3 m to 12 m below the surface. Under the influence of more normal events that entail wind-driven waves, rhodoliths on the edges of marine banks near the shore may roll away and onto a beach. Rhodolith debris crushed by wave surge and already on the beach may be transferred by wind farther inland to a coastal dune as, for example, on Isla Coronados near Loreto (Sewell et al., 2007).

Hurricanes represent a more extreme example of high-energy events, during which rhodoliths may be swept landward and stranded in a supra-

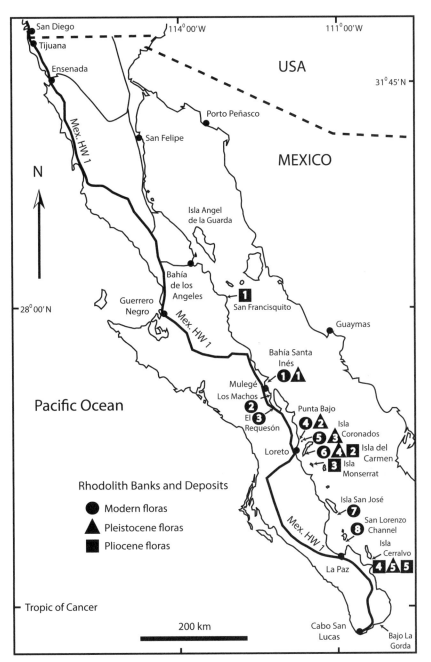

Fig. 5.4. Study sites for present-day and fossil rhodolith deposits in the Gulf of California.

tidal setting well above normal sea level where they expire and bleach under the sun. This outcome is documented both by recent storm deposits at Punta Bajo (Sewell et al., 2007) and Los Machos (Riosmena-Rodríguez et al., 2010) as well as fossil deposits on Isla Cerralvo (Johnson et al., 2012). The drop-off from marine shelves around banks and related islands in the Gulf of California typically is abrupt. It means that rhodoliths may be swept down slope during a hurricane to reach depths between 300 m and 700 m in complete darkness, as for example through the San Lorenzo Channel east of Isla Es-

píritu Santo near La Paz (Schlanger and Johnston, 1969). In total darkness without photosynthesis, the end result is death.

The ultimate high-energy event experienced in maritime lands is associated with the passage of tsunami waves (Fig. 5.3). The Gulf of California owes its existence to a tectonic regime that entails relatively frequent earthquakes. On the whole, these tend to be shallow in origin and low in intensity compared to earthquakes associated with plate subduction in marine trenches in places like

Chile and Japan. Tsunami deposits have yet to be confirmed anywhere around the Gulf of California. However, tsunami deposits that include rhodolith materials are recognized in other places such as the Cape Verde Islands more than 200 m above sea level (Ramalho et al., 2015).

Rhodolith Banks in the Gulf of California

Marked by circled numbers on the map in Figure 5.4, the distribution of active rhodolith banks in the Gulf of California is by no means comprehensive. Previous surveys identify as many as 60 widespread sites where living rhodoliths may be found (Steller and Foster, 1995; Riosmena-Rodríguez et al., 2010; Norris, 2014). The area offshore from Porto Peñasco in the upper Gulf of California is an example of one the more northern sites linked to reported rhodoliths. Bajo La Gorda at the southern tip of the Baja California peninsula is another. Herein, eight rhodolith banks briefly are described based on detailed survey work conducted as separate projects in those areas.

One of the largest rhodolith banks in the Gulf of Califorinia surrounds the Islas Santa Inés opposite Bahía Santa Inés near Punta Chivato (site #1). Shelf size may be appraised visually by the extent of turquoise-colored waters shown in an aerial view between the islands and the peninsular mainland (color insert, Plate 2). A 40-km² area was studied intensively on the basis of grab samples retrieved from the sea floor at depths up to 65 m (Halfar et al., 2012). Overall, carbonates account for 80% of the sediments retrieved from the samples. Whole rhodoliths and fragmented rhodoliths were found to be prevalent within a 17-km² area immediately around the Santa Inés islands. The bank is constrained within the 15-m contour for water depth, forming a substantial wave-dominated rhodolith bed. Morphological analysis revealed that 85% of the sampled rhodoliths are fruticose in form, 9% are lumpy, and 6% are foliose. The dominant fruticose rhodoliths are distributed across the entire bank, whereas lumpy forms are restricted to the east side (more exposed side) of the Santa Inés islands. The more fragile foliose rhodoliths are located leeward in the south part of the bank.

Southeast from Punta Chivato, several areas with high concentrations of rhodoliths occur within Bahía Concepción (Steller and Foster, 1995; Foster et al., 2007). Specific rhodolith banks summarized here are found at Los Machos (site #2) and El Requesón (site #3). The Los Machos bank covers a half-kilometer square area at depths from 2 m to 8 m below the surface (Foster et al., 2007). The dominant coralline red alga is *Lithothamnion muelleri*, but during the winter season the rhodoliths are invaded by the brown alga *Sargassum horridum*. Also on a seasonal basis, a sea urchin (*Arbacia incisa*) as well as various tunicates and marine worms are the most abundant invertebrates living among the rhodoliths. In terms of the rhodolith cryptofauna, 114 taxa were identified with an average of 40 taxa/individual rhodolith. Thirty-three species of fish were recognized on the rhodolith bank, dominated by grunts (*Haemulon maculicauda*) and porgies (*Calamus brachysomus*). The level of detailed survey work in this study (Foster et al., 2007) underscores the importance of the rhodolith ecosystem as a refuge for other marine organisms. Ashore at

Fig. 5.5. El Requesón tombolo in Bahía Concepción formed by rhodolith sand.

Fig. 5.6. Modern rhodoliths stranded on the beach near Punta Bajo. A) View with Isla Coronados in the distance. B) Close-up view (motor-oil container for scale = 25 cm long).

Los Machos, bleached rhodoliths in a supratidal setting attest to a standing event related to the passage of a major storm (Riosmena-Rodríguez et al., 2010).

The tombolo at El Requisón is composed of carbonate sediments derived mostly from crushed rhodolits (Fig. 5.5). The Spanish name for this locality refers to "cheese curds" that mimic the same texture as a heap of rhodoliths. Anecdotal history relates that masses of whole rhodoliths covered the beach during the aftermath of major storms. A detailed study based on long-term video surveillance in conjunction with the deployment of current meters, looked at the crescent-shaped rhodolith bank on the windward side of Isla El Requesón (Marrack, 1999). Rhodoliths from this wave-swept bed form dense concentrations 4 m to 12 m below the surface represented 89% by fruticose forms and 11% by foliose forms. In this case, both morphologies are attributed to *Lithophyllum margaritae*. Recorded at a depth of 5 m, small rhodoliths with a diameter of 2 cm start to roll under the influence of wind-driven waves when the water current reaches speeds between 25–30 cm/second. At higher velocities, between 30–35 cm/second, rhodoliths up to 3 cm in diameter start to roll away (Marrack, 1999).

Another phase of the same study using underwater video and current meters was performed at Punta Bajo (site #4) off the peninsular shores across from Isla Coronados near Loreto (Marrack, 1999). Typically from 2 cm to 6 cm in diameter, rhodoliths occur at a water depth of 12 m in the main channel between the mainland and the island. The rhodolilth bank at this locality covers 4.5 km2. The rhodoliths tend to be irregularly shaped and predominantly foliose in morphology mainly as a variant on *L. margaritae*. In a subsequent study at this site (Riosmena-Rodríguez et al., 2012), the species *Neogoniolithon trichotomum* also was found to be present. Maximum water movement associated with tidal currents was registered at 37 cm/second. Tagged rhodoliths were found to move only short distances back and forth. During a January test interval, northerly winds generated ocean swells up to 1.5 m high, but surface wave action had no effect on the channel rhodoliths.

Punta Bajo is a confirmed site where rhodoliths became stranded in a supra-tidal setting as a direct result of Hurricane Marty, which crossed the tip of the Baja California peninsula into the Gulf of California on October 22, 2003 and reached Loreto the next day. Visited less than three months afterwards, the supratidal deposit was found to cover an area roughly 35 m by 6 m parallel to the shore (Fig. 5.6A). The largest rhodolith observed in the deposit registered a diameter of 18 cm, but most were found to have diameters closer to 5 cm (Fig. 5.6B). Whereas daily tidal currents have little effect on the rhodolith bank offshore, storms of near hurricane intensity clearly have the capacity to move a significant volume of rhodoliths.

Shallow water around the west and south margins of Isla Coronados (site #5) entails an area covering 9 km² at water depths from 2 m to 6 m where rhodoliths are abundant. Bank size may be appraised visually by the extent of turquoise-colored waters shown in the aerial view around the island's west peninsula in the foreground (color insert, Plate 2). Rhodoliths with a sparsely branched fruticose morphology are most common, especially in the western embayment of Isla Coronados. The dominant species is *L. margaritae*, although *Neogoniolithon trichotomum* also is present (Riosmena-Rodríguez et al., 2012). The main beach on the west side of Isla Coronados is composed roughly 85% by the debris from crushed rhodoliths. Discussed further in Chapter 6, the adjacent sand dune registers a carbonate content of 87% derived from rhodolith detritus (Sewell et al., 2007).

Shallow water off the southern tip of Isla del Carmen (Fig. 5.4, site #6) covers the adjacent rhodolith bank at a depth less than 5 m over an area nearly 20 km² in size. Unlike rhodollith banks off Punta Bajo and Isla Coronados, the south Carmen bank features rhodoliths heavily dominated 70% by the species *Neogoniolithon trichotomum*, with the balance consisting of *L. margaritae* (Riosmena-Rodríguez et al., 2012). All the rhodoliths in this sector are fruticose in growth form, although 35% of the sampled rhodoliths also encrust pebbles. The extensive andesite rocky shore exposed around Isla del Carmen is the source for the pebbles. The beach at the southeast end of the island is 150 m long and spills seaward onto andesite basement rocks. Coarse sand-size bioclasts are composed more than 80% by the crushed and worn detritus of rhodoliths (Fig. 5.7).

Fig. 5.7. Rhodolith sand at seaward edge of beach on south tip of Isla del Carmen (pocket knife for scale = 9 cm).

Similar to Isla del Carmen, the shallow seabed off the southern tip of Isla San José and the adjacent Isla San Francisco (site #7) supports an immense rhodolith bank that covers an area of 45 km². The region was mapped in detail on the basis of an acoustic survey combined with the retrieval of sediment cores (Hetzinger et al., 2006). No distinction was drawn between live and dead rhodoliths, but the high concentration of rhodoliths at levels exceeding 60% of all carbonate materials on approximately 40% of the bank is an impressive finding. In addition to rhodoliths, other important contributors to the carbonate content were found to be clams, bryozoans, and corals. Data on species composition and growth morphologies were not divulged. Both of the large rhodolith banks associated with Isla del Carmen and Isla San José are developed on the leeward side of those islands in relatively sheltered settings.

South of Isla Espíritu Santo (Fig. 5.4, site 8), individual rhodoliths 3 cm to 6 cm in diameter are dominated by fruticose and foliose forms of *L. margaritae* living at a depth of 12 m (Marrack, 1999). Tidal current velocities are known to exceed 60 cm/second through the San Lorenzo Channel at this locality. During underwater video surveillance at this site, no rhodoliths were observed to shift location during tidal flows recorded at 37 cm/second (Marrack, 1999). It remains unclear whether or not rhodoliths move appreciably under a faster regime of tidal flow. Cores retrieved farther offshore well below the photic zone (Schlanger and Johnston, 1969) yielded rhodoliths presumably flushed through the San Lorenzo Channel into deeper water during major storms.

Fossil Examples and Paleoecology

As in fossil corals (Chapter 3) and fossil clams (Chapter 4), sea-level changes in conjunction with patterns of tectonic uplift during the past 5 million years are the chief agencies behind the accessibility of large rhodolith deposits located on land today. Examples of Pleistocene and Pliocene rhodolith beds are marked on the map by numbered triangles and numbered squares, respectively (Fig. 5.4).

Pleistocene relationships. Five examples are described briefly in this section, where rhodolith deposits of different grades (Fig. 5.3) occur in Pleistocene exposures. A 2.7-m thick deposit consisting of loose shells and abundant rhodoliths in rhodolith sand blankets a patch of ground roughly 0.75 km2 in area behind Playa La Palmita on Bahía Santa Inés (Fig. 5.4, site Δ#1). This extensive deposit represents a marine terrace assigned to the Upper Pleistocene Mulegé Formation that sits on a surface eroded in siltstone belonging to the Pliocene San Marcos Formation (Libbey and Johnson, 1997; Baarli et al., 2012). Landward, the deposit is backed by a low Pleistocene shoreline cut in Miocene andesite dating from the last inter-glacial epoch about 125,000 years ago (see Chapter 2). In addition to the formation's prolific rhodoliths, the upper part of the terrace features diverse nodules formed by separate clusters of barnacles, corals, bryozoans, and vermetid gastropods. Consistent with the term "rhodolith" these **macroids** are referred to as balanuliths, coralliths, bryoliths, and vermetuliths, respectively (Baarli et al., 2012). The assemblage shows that besides coralline red algae, many other organisms are able to form semi-spherical shapes capable of circum-rotary movement on the sea

Fig. 5.8. Pleistocene lagoon on Isla Coronados filled with bioclastic sand derived from crushed rhodoliths (person for scale, center right).

floor. Many of the non-algal macroids are equal in size to rhodoliths, commonly 3.5 cm to 4 cm in diameter. Essentially, the extensive terrace behind Playa La Palmita was a Late Pleistocene equivalent of today's Santa Inés rhodolith bank.

Sea cliffs near Punta Bajo that rise behind supratidal deposits with modern rhodoliths (Fig. 5.4, site #4) are composed of rhodolith limestone from the Upper Pleistocene (site Δ#2). These beds are significant, because they represent one of the few places in Baja California where fossil rhodoliths are preserved in a natural state close to original conditions that prevailed on a rhodolith bank. Johnson et al. (2009, their figs. 7.3A, B) illustrated different views of the fossil-rich formation. The basal part consists of fragmented rhodoliths reduced to coarse sand-size particles. In sharp contrast, the rhodolith sand is abruptly overlain by a 20-cm thick deposit of whole rhodoliths with an average diameter of 4 cm stacked three deep one above the other. The fossil rhodoliths are densely branched and fruticose in morphology. Whereas the layer of whole rhodoliths showing in situ preservation (as modeled in Figure 5.3) extends for only tens of meters, the greater limestone formation may be traced laterally for approximately one kilometer to and beyond Punta Bajo.

Pleistocene limestone composed of rhodoliths degraded to carbonate sand is widely ex-

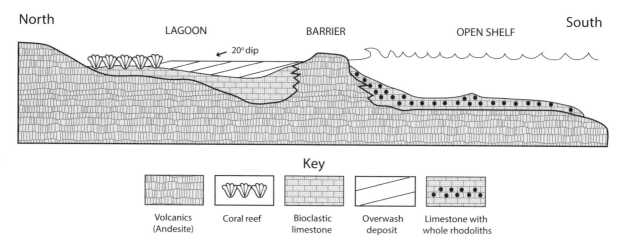

North
LAGOON   BARRIER   OPEN SHELF   South

← 20° dip

Key

Volcanics   Coral reef   Bioclastic   Overwash   Limestone with
(Andesite)   limestone   deposit   whole rhodoliths

Fig. 5.9. Geological cross section through the paleolagoon on Isla Coronados.

posed in Cañada Coronados on the south side of Isla Coronados (Fig. 5.4, site Δ#3). The sand surface was shingled by andesite cobbles due to runoff after a major storm event and the cobbles subsequently were colonized by a coral reef now elevated well above sea level (see Chapter 3). Radiometric dating of the corals is consistent with the last interglacial epoch sometime between 127,000 and 121,000 years ago. Covering an exposed area of 10,600 m² (Fig. 5.8), the site represents a former lagoon spread out at the base of a volcano last active about 600,000 years ago. The seaward margin of the lagoon is protected by a line of islets formed by Pliocene andesite with only a single narrow opening that allowed tidal flow in and out of the lagoon. A cross-section through the lagoon on a north-south axis (Fig. 5.9) shows the asymmetrical shape of the lagoon and the peculiar nature of rhodolith-

5.10. East face of Cañada Coronados showing rhodolith sand layers dipping northward (scale bar = 1 m).

sand layers that dip northward toward the volcano off the backside of the outer barrier formed by rocky islets. Viewed eastward perpendicular to the cross-section, an exposure in the sidewall of Cañada Coronados records multiple layers that dip 20° due north (Fig. 5.10). The layers represent overwash deposits that swept across the barrier on the lagoon's seaward margin (Ledesma-Vázquez et al., 2007). In part, this scenario conforms to the concept of an intertidal overwash deposit (Fig. 5.3), but as an extreme example owing to the high relief of the islets that normally sheltered the lagoon. High surf associated with the region's winter winds from the north were not a factor in this scenario, because the volcano protected the lagoon from wind and waves originating in that direction. Storms of near hurricane force that marched up the gulf from the south were the likely agency for the importation of rhodolith sand into the lagoon. Each layer (Fig. 5.10) appears to represent a separate storm event. It is significant to note that whole rhodoliths are found preserved in Pleistocene limestone outside the lagoon, but not within the lagoon, itself.

The south coast of Isla del Carmen has extensive rocky shores formed by low Pleistocene limestone fronting older Miocene andesite (see Chapter 2, Fig. 2.6). These deposits remain yet to be described in detail (site Δ#4), but are dominated by fossil rhodoliths and extensive rhodolith debris. At the southeast corner of the island, the same limestone formation brackets opposite ends of the magnificent, 150-m long beach (see book cover). The offshore rhodolith bank (Riosmena-Rodríguez

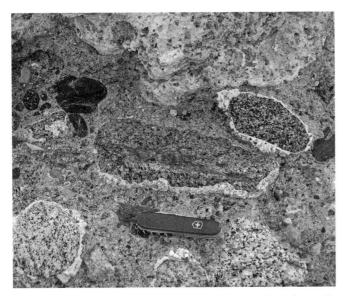

Fig 5. 11. Cobbles with encrusting rinds of coralline red algae at the base of a coral reef on Isla Cerralvo (pocket knife for scale = 9 cm).

et al., 2012), the beach, and the Pleistocene limestone all attest to a leeward setting in which rhodoliths prospered for thousands of years.

On Isla Cerralvo (Fig. 5.4, site Δ#5), the southwest corner of the island features a succession of coral reefs dated radiometrically to the last interglacial epoch 122,000 years ago (see Chapter 3). Each reef is anchored to a pavement of cobbles eroded from the island interior. The cobbles show different lithologies from granite to basalt to gneiss. Many of them are thinly encrusted by a rind of coralline red algae (Fig. 5.11). These rinds may not constitute true rhodoliths due to their relative thinness, and they tend to be thicker on one side as opposed to the other (Tierney and Johnson, 2012). Nonetheless, the clasts must have experienced some rolling action in order for the rinds to achieve total encrustation. Such movement appears to have taken place during long-shore drift from a tidewater delta located a short distance to the north.

Pliocene relationships. Five examples are described in this section, where rhodolith deposits of different grades (Fig. 5.3) are preserved in Pliocene rocks. The great Pliocene embayment surrounded by granite rocky shores at San Francisquito (see Chapter 2) has two entrances from the sea, both of which were functional during Pliocene time. The

north entrance features a rocky-shore overprinted by limestone that includes fossil rhodoliths (Fig. 5.4, site □#1). The dirt road leading to this locality ends at a cove 750 m long and 420 m wide at its present termination from Ensenada las Palomas to the north. The limestone abuts granite basement rocks that shed abundant conglomeratic cobbles and pebbles into the limestone (Johnson, 2014, p. 71). Many granite clasts are fully encrusted by coralline red algae with rinds up to 1.27 cm thick (Fig. 5.2C). The scenario meets the criteria of a relatively narrow tidal channel also influenced by wind-driven surf from the north. Based on the granitic conglomerate immediately adjacent to its source, a rocky-shore setting was in play, wherein wave action and turbulent flow eroded granite clasts and kept them moving in shallow-water much of the time. Elsewhere within the 10-km^2 basin, there is no evidence of rhodoliths having been present. Near the entrance to the channel, however, wave action promoted the growth of rhodoliths nucleated around the pebbles.

The largest rhodolith deposit in the Gulf of California is found at Arroyo Blanco on the southeast coast of Isla del Carmen (site □#2). The location was investigated by Eros et al. (2006), who logged a 157-m thick section of inter-bedded limestone and thin conglomerate wedges that span much of the Pliocene and Pleistocene. Relative

Fig. 5.12. Mouth of Arroyo Blanco on the east coast of Isla del Carmen with massive beds of rhodolith debris.

dating was based on the occurrence of overlapping range zones defined by several species of fossil bivalves and echinoderms related to the stratigraphic scheme devised by Durham (1950). The vertical range of key index fossils recorded in the lower walls of the canyon near the coast include a large pecten (*Patinopecten bakeri)* that has long been extinct and a stubby sea urchin (*Clypeaster bowersi*), also now extinct in the Gulf of California. Massive limestone formations exposed at the mouth of the canyon (Fig. 5.12) are entirely dominated by rhodolith debris with whole rhodoliths only rarely evident.

As the preserved part of the Pliocene basin has a surface area of 3.3 km² and the portion filled with materials derived from rhodoliths is estimated to account for 64% of the basin's volume, it is calculated that roughly 250,000 cubic meters of rhodolith material are buried against the side of a former rocky shore at Arroyo Blanco. Experimental data show that it takes 8,640 whole rhodoliths (5 cm in diameter) to fill a one-meter-square box, based on a ratio using a much smaller box (Sewell et al., 2007). After reducing rhodoliths into carbonate sand having a grain size of 2.38 mm (or less), proportional arithmetic indicates that it would take 16,265 whole rhodoliths to produce an equivalent amount of rhodolith sand to fill a one-meter-square box. Thus, it may be assumed that somewhat more than 4 billion rhodoliths went into the accumulation of rhodolith limestone in the Arroyo Blanco basin over several million years.

A similar large basin with extensive rhodolith limestone exists on the west side of Isla del Carmen at Bahía Marquer. At 3.42 km², the size of that basin covers an area slightly larger than the Arroyo Blanco basin. Coastal cliffs that cross the basin vary in thickness from 17 m in the north to 20 m in the south. The lower 10 m of the succession visible at the coast is enriched in rhodoliths roughly 60% by volume (Johnson et al., 2009). The Bahía Marquer basin remains yet to be studied in greater detail, and hence comparative data regarding the cumulative input of rhodolith material remains to be determined.

Considering local tectonics, the most intriguing rhodolith limestone of Pliocene age in the Gulf of California is on Isla Monserrat (Fig. 5.4, site □#3). Difficult to access, the site consists of a 3-m thick caprock that covers an area about a third of a square kilometer near the island's center (Johnson, 2014). The plateau sits at an elevation 193 m above sea level, resting on tilted layers of andesite that date from the Miocene. At the center of the plateau, abundant whole rhodoliths are weathered free from the limestone. More research remains to be done, but at least three genera of coralline red algae are represented by the Monserrat rhodoliths (*Lithothamnion*, *Sporolithon*, and *Lithophyllum*). The most abundant and largest rhodoliths (up to 4 cm in diameter) are thinly branched fruticose types. Less common but almost as large are the lumpy forms with thick branches and stubby tips. This particular patch of limestone is isolated by erosion, but other patches dot the roof of the island.

One kilometer farther to the southeast, Pliocene limestone abuts laterally against Miocene andesite that signifies a former rocky shore (see Chapter 2). Thus, the limestone patch at the middle of the island has the bearing of a rhodolith bank in the context of a paleoshore much like the present-day Isla Santa Inés bank associated with Bahía Santa Inés on the peninsular mainland. The amount of uplift shown by limestone formations on Isla Monserrat is impressive, and attests to ongoing geological stresses in the Gulf of California due largely to transtensional tectonics (see Chapter 1).

Fig. 5.13. Pliocene strata at Paredones Blancos (White Cliffs) on the west coast of Isla Cerralvo.

Two localities with Pliocene limestone dominated by rhodoliths and rhodolith debris occur on the opposite coasts of Isla Cerralvo (Fig. 5.4). Part of the west coast (site □#4) features a distinctive landmark called Paredones Blancos (white cliffs). The cliff line, which extends for a distance of 0.75 km, consists of a 10-m thick unit of rhodolith limestone bracketed below and above by equally thick units of unfossiliferous conglomerate (Emhoff et al., 2012). Although readily accessible by boat, the cliffs are steep and unstable (Fig. 5.13). Huge boulders of the rhodolith limestone litter the beach, and these are easy to reach and to sample within their stratigraphic context. Not surprisingly, thin sections from samples collected through the limestone sequence show that the middle part of the limestone unit is least contaminated by clastic sediments. Few whole rhodoliths are present in the limestone, which consists overwhelmingly of finely crushed rhodolith debris. Overall, the succession of conglomerate and limestone units is interpreted as due to fluctuations in Pliocene sea level (Emhoff et al., 2012). The volume of rhodolith limestone at this locality is difficult to calculate, due to extensive faulting.

Fig. 5.14. Pliocene rhodoliths at Los Carillos on the southeast coast of Isla Cerralvo.

The southeast shore of Isla Cerralvo at Los Carillos is the location that features a major standing event of whole rhodoliths preserved as part of a Pliocene sequence (site □ #5). The locality includes a former granite shoreline with associated conglomerate that contains fossil gastropods (*Nerita scabricosta*) characteristic of a rocky-shore setting (see Chapter 2). Extensive patches of limestone with abundant rhodoliths reside among small granite boulders that poke through the limestone (Fig. 5.14). The rhodoliths are mostly 3.5 cm in diameter and occur in lenses up to 15 cm thick spread over an area covering 150 m^2 (Johnson et al., 2012; Johnson, 2014). The Pliocene scenario is strikingly similar to the modern rhodolith stranding event associated with Hurricane Marty at Punta Bajo north of Loreto (compare with Fig. 5.6A).

In summary, four locations with Pleistocene rhodolith deposits described herein occur near modern rhodolith banks of considerable size off Bahía Santa Inés (site #1), at Punta Bajo (site #4), the southeast shores of Isla Coronados (site #5), and the south coast of Isla del Carmen (site #6). However, none of the five locations with Pliocene rhodolith deposits in this summary occur near modern rhodolith localities of any size. For example, a systematic search for modern rhodolith beds around Isla Cerralvo resulted in the location of a single spot off the north end of the island around a seamount called El Bajo de la Reina (Backus et al., 2012). The spot is isolated from the island shelf by a deep channel, and is immune to the large volume of clastic sediments eroded from the island due to frequent storm action. Other Pliocene rhodolith beds have undergone substantial tectonic uplift on islands such as Monserrat and Carmen, where the adjacent insular shelves are now quite narrow and drop off rapidly into much deeper water. Even so, the Pliocene limestone beds preserved on those gulf islands demonstrate the monumental scale of rhodolith production dating back millions of years in geologic time.

Chapter 6

Sandy Beaches and Coastal Dunes

Introduction

Baja California gulf shores exhibit hundreds of beaches, some of which are pocket beaches only a few tens of meters wide isolated between imposing rocky cliffs. Other beaches offer open expanses that appear unlimited to the eye. In the upper gulf north of San Felipe, beaches connected to the coastal highway by dirt tracks merge together as a single, 8-km strand between Playa El Paraiso and Playa El Blanco. From San Felipe south to Estrella, the beach is continuous for 14 km. At the south end of the gulf between Los Frailes and La Fortuna, a ribbon of sand hugs the low coast uninterrupted for more than 20 km. Likewise, there are many coastal dunes dispersed at places adjacent to the gulf coast. Mapped using satellite images, 84 coastal dunes are registered along the peninsular coast and on associated islands (Backus and Johnson, 2009). Most are low, seldom rising more than 2 m above the adjacent shore. In contrast, the dunes around Cerro El Gallo near Mulegé reach a maximum height of 21 m (Skudder et al., 2006

This chapter examines sandy beaches with attached dunes, where onshore winds have a long history of denuding the finer fraction of beach sand for transport inland off the beach. The preceding sections on rocky shores (Chapter 2), clam flats (Chapter 4), and rhodolith banks (Chapter 5) are relevant, because shore orientation and the concentration of calcareous shells and algae in preferred life zones is vital to the enrichment of beaches and dunes in the middle and northern parts of the gulf. A kind of natural conveyor belt operates through wind and waves to bring biological refuse to the foreshore and shore, where the mechanical work of the surf reduces those materials to organic sand. Pocket beaches in the north are framed by cliffs composed of andesite and basalt that also contribute pebbles and dark mineral grains to the sand.

Where the peninsular coast is dominated by granite southward from La Ventana and Isla Cerralvo, carbonate sand is scarce. Including a bend in the coast at Punta Arena, the beach sand over much of that region is highly enriched in quartz grains derived from the breakdown of granite. Marine life is no less abundant in the lower gulf, but other factors related to weather patterns appear to play a crucial role in the disparity.

Aside from variations in the biological fabric of beach and dune sand within the Gulf of California, exploration of the foreshore, beach, and related dunes offers insights on processes that operate everywhere to shape the coast. Perceptions are gained through what can be felt by our feet in wadding out over sand bars and by our hands in crawling across beaches, as well as climbing up and down adjacent dunes. The result is a better understanding of the relationship between the physical world influenced by wind, water, and rock in conjunction with the biological world that presses against the shore.

Abatement of Wind and Wave Energy

Mornings on the gulf most often are a time of calm when the water surface is dead flat. Anyone accustomed to boating expects that white caps may appear on the crests of waves that grow increasingly agitated as the day wears on. Under steady winds that blow for days at a time, sea swells take on an added dimension to stir gulf waters into uniform motion at the surface. During the winter months from November to April, high-pressure fronts in the far north are funneled down the axis of the gulf to send bursts of wind and wind-driven waves southward. The pattern may reverse itself during the summer months from May to October, when lighter winds sometimes push waves northward (see Chapter 1). Waves carry energy that is transmitted to the sea floor with variable effect, based on the size of the waves and the distance between wave crests (or swells). Under persistent winds with swells separated by a distance of 30 m, for example, some of the energy produced by the passing swell touches the sea floor at a depth of 15 m. As a rule of thumb, the vertical depth of water in which a wave "feels bottom" is calculated as equivalent to one-half the horizontal distance between adjacent swells.

Energy transferred to the sea floor stands to increase as waves reach shallower water on approach to land. Swells change to turbulent surf when water molecules at the crest of a swell surge forward at a faster speed than the movement of water molecules below that are slowed by friction with the sea bottom. Even on a relatively quiet day under light winds, beach watchers speak of waves that curl and break as they arrive close to shore. Over time, large-scale patterns of sedimentation and the development of sedimentary rock formations with affinities to windward and leeward settings owe their origin to these basic forces of atmospheric and marine circulation in the Gulf of California. Based on experiences during calm and stormy days, alike, the dedicated beach watcher is compelled to consider the following question. What becomes of all the energy brought by the wind and waves from a direction that collides head-on with islands and peninsular shores in their direct path?

Waves strike both head-on against an obstruction to exert maximum force, but they also bend to pass around and impart less force behind the same obstruction. An outstanding place to visualize the effects of **wave refraction** is the Punta Chivato promontory, located midway between Santa Rosalía and Mulegé. The Mexican topographic map for the area applies the Spanish word atravesada to the promontory, which translates into English as a passage or "cross-piece." The Punta Chivato area features approximately 25 km² that protrude as a headland eastward into the Gulf of California and offers a stretch of exposed coast facing directly to the north. The topographic map also informs

that the body of water immediately north of Punta Chivato is called the Ensenada El Muerto.

Fixed as a reference point diagonal to Isla San Marcos and the Punta Chivato promontory, the 50-fathom isobath (equivalent to a water depth of 90 m) represents a break on the sea floor over which the winter winds and waves pass on the way toward land (Fig. 6.1). Southward beyond that line, waves reach increasingly shallow water. Each square or rectangle in the grid that straddles the 50-fathom isobath signifies an equivalent unit of energy. There is no change in the amount of energy delivered to the north shore at Punta Chivato, but sea swells passing Punta Chivato to the east are deflected in a westward direction. Units of energy are deformed as rectangles that are two to four times as large as the standard squares elsewhere in the grid on approach to Playa La Palmita and Punta Mapachito in Bahía Santa Inés. The grid provides a graphic way to show how waves impacting the south side of the Punta Chivato promontory are four times less energetic as those that assault the north side. On closer examination, it can be seen that Isla San Marcos shields the coastline around El Rincón to some degree, although wave energy is not dissipated nearly as much as at Punta Mapachito in the south (Fig. 6.1). Overall, the north-facing beaches

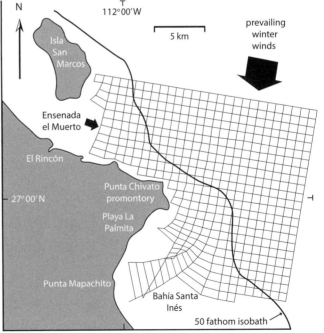

Fig. 6.1. Computer simulation showing wave refraction at Punta Chivato.

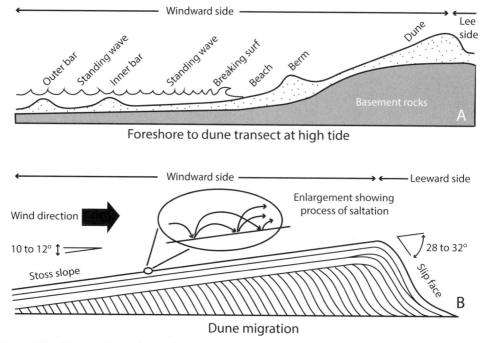

Fig. 6.2. Dynamics of wind and waves on north-facing shores. A) Foreshore to dune profile, B) Transverse dune profile.

of Punta Chivato have a reputation for being both violent and deadly during the winter months.

A profile across the foreshore, beach, and coastal dunes on a north-south transect (Fig. 6.2A) is characteristic for the north shore of Punta Chivato during high tide. The beach and its related berm are situated in the middle of the profile. The berm represents an accumulation of sand during the confluence of storms during the highest tides that occur twice a year during the equinoxes when the gravitational pull on water is enhanced by alignment of the sun, earth, and moon. Spring tides always occur twice a month in concert with the full moon, but those tides are made even more extreme during the equinoxes. The transect shows only two offshore sand bars, although during a low tide it is possible to wade off shore and experience the swells and surf that cross as many as four sand bars parallel to the shore with the outermost bar located 120 m out (Johnson, 2002, p. 105). Essentially, the water that crashes onto a beach under the direct advance of waves has no place to go except to retreat offshore as an outgoing wave. Sand bars form as a result of a phenomenon called a **standing wave**. Especially during a low tide, it is clear to see that waves expend energy crossing the sand bars

that stand in their way. However, when a retreating wave collides with an advancing wave, the result is said to represent a **nodal point** where the energy is largely nullified.

At a nodal point midway between two sand bars, it is possible to feel a kind of whipsaw action about your feet as small bits of shell and coarse sand shift rapidly back and forth on the floor of the trough. The sand bars represent **antinodal points**, where one can feel the passage of each swell, but the main sensation is that of bobbing up and down. Sand bars build at the antinodes, but they may shift in position forward and backward both under the influence of stronger tides and the effect of stronger wind-driven waves. Ultimately, all the water pushed landward by the waves must retreat somewhere with an equal efficiency that circumvents the several standing waves located at nodal points between sand bars. The result is the creation of strong rip currents that converge from opposite directions parallel to the beach to break outward perpendicular to the beach. Except during slack water when the tides turn, the power of the rip tides is formidable.

The sandy shores on the Ensenada el Muerto exhibit one of the finest carbonate beaches in all of Baja California. Great patches cover the beach with bits of shells worn into smooth disks the size of coins. To hold one of the larger shell bits is to experience the silky smoothness of a flattened worry bead. Some dark pebbles of black basalt and red andesite litter the beach, but overall the white beach sand is composed of shelly material largely derived from mollusks. During high tides, the shell shingle is subjected to the pounding surf, which functions to grind shell pieces into smaller and smaller bits. Over time, such action results in the production of very fine sand particles derived from the shells. This particular fraction of the beach sand is vulnerable to removal by the same winds that push the waves onto the beach. The transfer of beach materials inland by the wind is called **beach deflation**.

The longitudinal profile of a dune rising behind a beach berm (Fig. 6.2B) is typical of many north-facing shores along the Gulf of California. Dunes exhibit two slopes with opposite orientations and very different angles of inclination. The windward incline is the **stoss slope** that rises upward like a ramp at a low angle generally between 10° to 12°. The leeward slope is much steeper, falling at an angle often between 28° and 32°. The leeward slope is called the **slip face** and the dune's crest represents the leading edge where sand grains are pushed over the top by the wind. Air directly behind a dune is becalmed and the sand that tumbles down the slip face accumulates as layers tilted at the **angle of repose**. This angle reflects the steepest possible slope on which sand (or larger particles like the rock scree on the side of a mountain) may reside naturally under the force of gravity. Any disturbance on the slip face of a dune, such as an attempted climb, provokes a minor sand avalanche.

Sand grains climb the stoss slope to the crest of a dune by **saltation** (Fig. 6.2B). When an individual grain of sand carried by the wind lands on the stoss slope, it strikes one or more grains at rest and transfers its energy to those grains. The effect of the collision is to eject those grains into the air, where the wind carries them farther up the stoss slope in an arc-shaped trajectory. The term "saltation" is a derivation of the Latin word saltator for

a dancer or one who hops. Hence, it may be said that sand grains hop or jump forward to the leading edge of a dune. Silica sand grains from a dune setting are highly spherical in shape, but carbonate grains derived from mollusks tend to be more plate-like in shape due to the prismatic microstructure common to many clamshells. Such carbonate grains may be more aerodynamic and capable of longer flights, or jumps. The distance above the surface of a stoss slope where dune sand is normally carried by the wind is in the range of 20 to 40 cm. Walking barefoot up the stoss slope on a windy day, one feels the sting of wind-blown sand around the ankles and lower legs.

Effect of Wind on Coastal Vegetation

Over time, dunes that penetrate well inland become stabilized by vegetation that includes grass, shrubs, and even small trees. The wind, however, has a constant effect on dune vegetation and those plants that grow on the rocky landscape surrounding dunes. For example, the shifting wind has the effect of sweeping a rooted grass plant back and forth like a broom where the plant is pressed flat on the ground. The effect is to leave a pattern inscribed in the sand that accentuates irregularities in the plant's stem as a wind rose (Plate 2). In this case, the grass behaves like the waving windsock at an airport to show the changing direction of the wind. Under the right conditions, such a pattern may be preserved in dune rocks. Larger shrubs, such as the leather plant (*Jatropha cuneata*) become permanently deformed by the wind and hug the ground in a down-wind direction. Even trees like the copal (*Bursera hindsinia*) become permanently bowed by the wind and may stretch out on the ground for several meters from the windward side of the tree trunk to the tip of leaves on branches extended in the leeward direction.

This kind of deformation is called a ***krummholz***, from the German for "twisted wood." Plants are pruned indirectly by the wind, with plant tissues exposed to an aerosol of misty salt that stunts upward growth. In order to survive in this setting, plants are forced to grow downwind close to the ground such that branches and leaves are protected in the shelter of roots and stem. Leather plants and copal trees, together with several other kinds of shrubs were measured with respect to their growth orientation on the windward face of the Punta Chivato promontory (Russell and Johnson, 2000). When plotted in a rose diagram, the fixed orientations measured from a sample of 168 plants clustered together with an average preferred direction pointing nearly due south. Plants that thrive on the leeward slope of the Punta Chivato promontory are perfectly normal with healthy growth that points upward. Salt aerosols fail to reach the south side due to shelter from the wind provided by the promontory. Based on pencil cores drawn from some of the larger copal trees on the windward face of the promontory, growth bands indicate that trees have grown under stunted conditions for up to 45 years (Russell and Johnson, 2000, p. 715).

Attached Beaches and Dunes in the Gulf of California

Numbered circles on the map in Figure 6.3 mark the location of many outstanding examples of co-joined beaches and coastal dunes. There are scores of possible sites for study. It is worth noting that the extensive treatment on this topic by Backus and Johnson (2009) plots wind indicators such as unidirectional ripple marks from as many as 74 dunes at localities as far apart as San Felipe in the north and La Fortuna in the south. The resulting rose diagram resembles that based on plant orientations from Punta Chivato, but represents a wind field over the entire Gulf of California showing the nearly due south migration of sand dunes by incremental saltation.

Research on carbonate dunes in the Gulf of California began with the study by Ives (1959) on dune fields from the Sonoran coast (Fig. 6.3, site #1). It was discovered that sand dunes close to Sonoran beaches are composed more than 70% in volume by shell bits ranging from 0.01 to 2 mm in diameter. By comparison, sand dunes across the border in Yuma, Arizona were found to be com-

posed 90% by silica sand. The Yuma dunes have a source from sand carried by the Colorado River. The immediate source for the Sonoran dunes with a high concentration of shelly sand are the clam flats in the adjoining gulf. It was Ives (1959) who associated the winter norte winds with beach deflation and the development of carbonate dunes.

Previously cited as a locality in Chapter 3 on clam flats, the north-facing beach near Bahía de los Angeles (site #2) is the source for the extensive berm deposits dominated by clamshells. Sand dunes extend far inland across the floor of the Valle las Flores to the south. To the extent that carbonate sand is present in these dunes, the immediate source is the southern part of the adjoining bay. Playa San Rafael is situated some 57 km southeast from Bahía de los Angles (site #3), where the wide beach is accessible by road. The beach occupies a bend in the shore open to the north and reveals a worn berm and broad inter-tidal flat (Fig. 6.4). The dunes that rise behind the beach reach a height 10 m above sea level.

Dunes with adjoining beaches extend eastwards from El Rincón (site #4) along the shores of the Ensenada El Muerto (site #5) to Punta Chivato (site #6) interspersed over a distance of 12 km. The dunes at El Rincón cover an area of 2.5 km² and rise to a height of 18 m above sea level. The beaches and associated dunes along the middle shores of Ensenada El Muerto (Fig. 6.5) consist of sand that on average contains more than 50% of materials derived from mollusks (Russell and Johnson, 2000), with samples at beach level that approach 80% in bulk concentration of mollusk sand. The dunes behind the beach are not so impressive for their size as they are for the way that dune sand drapes a prominent vertical step in andesite elevated 10 m above the beach. Maximum elevation of dune sand at this locality is 60 m above sea level. Maximum wind speed measured during field studies was 8 m per second. Air-borne sand grains were felt at a distance 1.75 m above the ground (Russell and Johnson, 2000, p. 717). At Playa Cerotito closer to Punta Chivato (site #6), a thin layer of dune sand is plastered on the north face of a prominent andesite

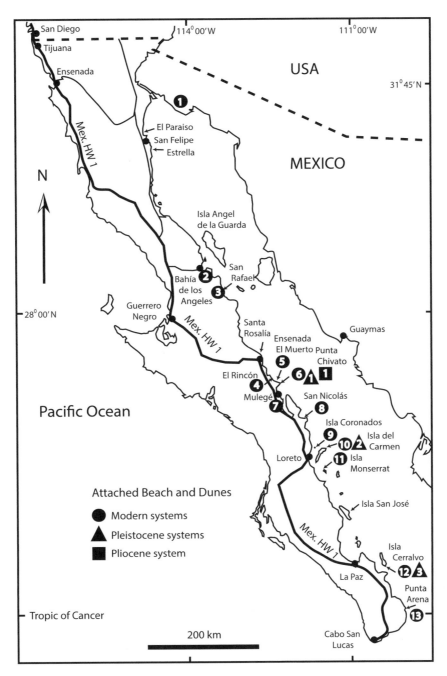

Fig. 6.3.   Study sites for present-day beach and dune systems in the Gulf of California including geological examples.

dome that rises to an elevation of 40 m above sea level (Fig. 6.6). Sand samples collected from beach level to the top of the dome show little variation in grain size, but the concentration of mollusk sand ranges between 30% and 40%. Lithic grains of volcanic sand are more common on the beach (36%) but decline toward the top of the dome (12%). The relationship is due likely to the higher density of volcanic grains compared to shelly grains, which makes them harder to carry by the wind.

Dune fields near Mulegé around Cerro El Gallo cover nearly 2 km² in surface area, and penetrate as far inland as 1.75 km from an east-west beach (Fig. 6.3, site #7). The leading edge of the dunes is impressive for its height (Fig. 6.7). Compared to the dunes around Punta Chivato, however, they register a much lower content of shelly material. Dune sand nearest to the beach has a content of about 15% carbonate grains, whereas the level of enrichment falls to 5% in samples tested farthest

Fig. 6.4. Beach and dunes at Playa San Rafael.

Fig. 6.6. Beach and dunes off Playa Cerotito.

Fig. 6.5. Beach and dunes on the Ensenada El Muerto near Punta Chivato.

Fig. 6.7. Dunes at Cerro El Gallo near Mulegé.

from the beach (Skudder et al., 2006). The input of volcanic grains at this site comes from two local sources, the most immediate being the andesite rocks of Cerro El Gallo on the east edge of the dune field and the small Isla El Gallo located less than 100 m offshore to the north. A more significant source of volcanic sediment is the Mulegé estuary that empties into the Gulf of California 5 km northwest of Cerro El Gallo. The Mulegé River drains an area covering about 250 km² in a landscape dominated by volcanic rocks. Long-shore currents pushed by the winter winds are fully capable of delivering eroded river sands from the estuary to the beach and adjacent coastal dunes southeast of Mulegé.

Near the village of San Nicolás some 65 km southeast of Mulegé (Fig. 6.3, site #8), a north-facing beach with abutting dunes extends for 3.25 km in length. The surface area of the San Nicolás dunes is about the same as those near Mulegé, but

they hold less volume and are less active due to better stabilization by vegetation. The active part of the dune field is limited to a narrow strip parallel to the beach, although a thin cover of sand extends inland as much as one kilometer to reach an elevation 40 m above sea level (Skudder et al., 2006). The stoss slope on these dunes is much eroded and reconfigured by storm wave. At the far end of the beach, the most active dunes exhibit a slip face with a maximum drop of 7 m. In contrast to the Mulegé dunes, those near San Nicolás are more highly enriched in shelly materials that register up to 50% in volume from samples collected closest to the beach and 25% more inland. At low tide, the adjacent beach has an exposure roughly 30 m wide. Beach crabs are present and colorful inhabitants (Plate 2). Chocolate shells (*M. squalida*) disarticulated as separate valves (Fig. 6.8A) commonly occur as debris on the lower beach. Broken shells derived mostly from the same species (Fig. 6.8B) cover the

Fig. 6.8. Shell debris from a beach near San Nicolás. A) Disarticulated chocolate shells from the lower beach, B) Shell hash from the upper beach.

upper beach. Dune sand cannot be tied explicitly to this species, but the prismatic microstructure of clamshells is a distinctive trait easily recognized in samples viewed in thin sections under the microscope.

Box experiments performed to test for the number of articulated chocolate shells needed to yield a measurable amount of finely crushed material can be checked with the volume of dune sand that yields a given concentration of shelly grains. By this method it is possible to crudely estimate the size of bivalve populations living offshore required to supply a given beach and associated dune field with shelly materials over time. With an estimated volume of 626,400 cubic meters, the San Nicolás dunes reflect an average level of carbonate enrich-

ment requiring the input of approximately 1.36 billion clams (Skudder et al, 2006). Although the Cerro El Gallo dune field near Mulegé registers a much lower level of enrichment from shelly materials, it holds a much larger volume of sand estimated at 11 million cubic meters (Backus and Johnson, 2009). According to calculations using the chocolate shell, the number of clams from the offshore Mulegé flats contributed to the Cerro El Gallo dune field over time amounts to approximately 6.2 billion clams. Similar calculations have yet to be made for the dune fields at El Rincón and Punta Chivato, but the abundance of chocolate shells on the Ensenada El Muerto is sufficient to the task.

Introduced in Chapter 5 on rhodolith banks, Isla Coronados features an array of recent and Pleistocene deposits substantially enriched by carbonate debris derived from the coralline red algae. Prominent among them is the beautiful beach and attached dune field on the west side of the island (Fig. 6.3, site #9) much enjoyed by people on day trips from nearby Loreto. The beach follows a crescent-shaped path that extends for 330 m. Small mollusk shells commonly contribute to the beach's composition, but crushed rhodoliths are the primary component. Wave refraction across an offshore rhodolith bank to the west is the source for these sediments. The dune field adjacent to the beach has a surface area of 15,000 m^2 with a leading edge at a slip face advancing to the southeast. Maximum dune height is 4 m and it is calculated that the overall dune volume amounts to 30,000 cubic meters enriched 87% by sand derived from rhodoliths (Sewell et al., 2007). In comparison, shell fragments derived from crushed mollusks account for only about 7% of the material sampled from the dune.

Box experiments performed to test for the number of whole rhodoliths (average diameter of 5 cm) needed to yield a measurable amount of sand-size particles ($\leq$ 2.38 mm) can be multiplied by the volume of dune sand enriched by rhodolith debris. By this method, it was calculated that the Isla Coronados dune field holds a volume of sand at an average level of carbonate enrichment requiring the input of nearly one-half billion whole rhodoliths (Sewell et al., 2007). So far, no other beach or related dune field has been tested for raw input of rhodolith materials, but other localities exist and should be tested.

A cove at the far northeast end of Isla del Carmen (Fig. 6.3, site #10) provides an outstanding example of a pocket beach with adjoining dune that fits the profiles illustrated in Figure 6.2. Barely 60 m wide, the beach is framed by limestone sea cliffs that extend seaward for 350 m on opposite sides of the narrow inlet. The associated dune exhibits a north-facing stoss slope 185 m in length with a leading edge that stands 17.5 m above sea level. Dune sand at this location has an estimated volume of 75,000 cubic meters, and it was found to be enriched 68% by carbonate grains derived from clamshells (Backus and Johnson, 2009). Following the same computations applied to the dune field near San Nicolás, the number of articulated bivalves contributed to this small dune is equivalent to 290 million chocolate clams.

Situated 45 km southeast of Loreto, the north-facing shores of Isla Monserrat (Fig. 6.3, site #11) provide more examples of beaches with related dunes. In this case, a small patch of Pliocene limestone breaks a one-kilometer long stretch of shoreline otherwise dominated by sandy beaches that sit side-by-side like twin pocket beaches (Fig. 6.9). Pavements with worn bits of shell (Plate 2) cover parts of the beaches with materials derived mostly from the epifaunal bittersweet shell (*Glycymeris*

maculata) and infaunal chocolate shell (*M. squalida*). Low dunes behind the beaches reach an elevation 22 m above sea level and hold an estimated volume of 30,000 cubic meters. Dune sand from this location was tested for composition and found to be enriched 42% by carbonate grains with affinities to clamshells (Backus and Johnson, 2009). Application of the same calculations used for the San Nicolás dune field yields an estimate for the number of articulated bivalves contributed to the Isla Monserrat dunes as equivalent to about 70 million chocolate clams.

Beach and dune sands near the southwest light tower on Isla Cerralvo (Fig. 6.3, site #12), form a great bulge that points westward off the island toward La Ventana across the intervening channel on the peninsular mainland. The beach follows a bend through a 60° angle that measures 800 m on each side with an exposed width between 70 and 80 m during low tide (Fig. 6.10). A triangular-shaped dune defined by the beaches on the seaward margin and blocked by an abrupt rocky slope at the landward edge covers an area of 170,000 m² to achieve a maximum inland elevation of 15 m. This feature represents one of the largest active fan deltas on Isla Cerralvo, among 39 such deltas surveyed around the island's periphery (Backus et al., 2012). Emerging at the far north end of the beach, a normally dry stream bed leads inland for 3 km with an average slope of 17°. The valley surrounding the arroyo defines a drainage basin with an area of nearly 3.5 km². During episodic rainstorms, granitic material that include abundant sand, cobbles, and small

Fig. 6.9. Beach and dunes on the north shore of Isla Monserrat.

Fig. 6.10. Beach and dunes on the southwest shore of Isla Cerralvo.

boulders are fed to the delta by stream erosion. The delta sediments are sorted and removed by a vigorous long-shore current along the coast during the winter season, stimulated by stiff winds out of the north. The same winds deflate the north beach, sending sand onto the stoss slope of the associated dune.

A similar example occurs at Punta Arena, where a great bend on the peninsular coast appears 12 km southeast of La Ribera (Fig. 6.3, site #13). The angle of the bend amounts to about 80°, defined by the intersection two beaches each 5 km in length that meet at Punta Arena. The dune inscribed between the beaches covers an area of 9,780 m² with an estimated volume of 489,000 cubic meters. Like the beaches of Isla Cerralvo, the content of the Punta Arena beaches is dominated by well-sorted sand composed almost entirely of silica grains derived from the decomposition of granite. A local source for the silica sand at Punta Arena is the river mouth that opens onto the Gulf of California at La Ribera. The region's former and present geothermal springs (see also Chapter 10) have been and remain influential in degrading granite in parts of the regional terrain from La Ventana southward to Cabo San Lucas. With stream erosion due to higher than average rainfall amounts in the southern gulf region associated with episodic storms, silica crystals are readily separated from weathered granite and transported downstream to build beaches and coastal dunes.

Paleoecology of Former Beaches and Dunes

Geological evidence for former beaches and related dunes is not as widespread as for many other ecosystems in the Gulf of California, but does exist and provides supporting data on patterns of atmospheric and marine circulation during the gulf's paleoecological history. Three examples of Pleistocene beaches and dunes are marked on the map by numbered triangles and a single area is identified by a numbered square for a Pliocene dune field (Fig. 6.3).

Fig. 6.11. Pleistocene beach deposit on Playa Cerotito near Punta Chivato.

Pleistocene relationships. A profile of an Upper Pleistocene beach is exposed by erosion at the seaward edge of the 10-m marine terrace behind Playa Cerotito on the east end of the Punta Chivato promontory (Fig. 6.3, site ∆#1). In this case, the cross section through a thickness of 1.5 m reveals distinctive bedding in a striking pattern of thin centimeter-scale layers dominated alternately by dark andesite pebbles and white shell fragments (Fig. 6.11). Parts of this beach deposit abut a former cliff line. The rocky shore forms the exposed base of a large andesite dome that was extruded during the Miocene. The Pleistocene beach sequence is judged to be somewhat older than 125,000 years, due to the fact that the deposit's upper surface marks the 10-m marine terrace occupied during the last interglacial epoch when sea level was higher than today.

The alternation of black and white layers implies seasonal variations in storms arriving from the north. Wave activity of greater intensity was required to emplace an apron of relatively heavy andesite pebbles across the beach during stronger winter storms. More normal wave activity was likely to entail deposition of sand bits derived from mollusk shells during the balance of the year. No dune rocks are preserved at this locality, but the andesite dome was certain to have served as the same kind of windbreak that remains in effect, today.

The north end of Isla del Carmen features extensive Pleistocene dune rocks now exposed as massive sea cliffs that rise vertically 70 m in thickness from the water to reveal at least five different stages of overlapping dune development (Johnson and Backus, 2009, p. 143). Anderson (1950, p. 20) was the first to describe these deposits, which he characterized as "cross-bedded calcareous sandstone." Previously introduced in this chapter as site #10, the most accessible spot from which to explore the Pleistocene dune rocks on Isla del Carmen is located the same place where a modern-day pocket beach and attached dune occupy a narrow cove (Fig. 6.3, site Δ#2). The sides of the inlet are composed of cross-bedded calcareous sandstone (Fig. 6.12). Fossil land snails (air-breathing gastropods) are preserved within these strata (Fig. 6.13), which clearly confirms a terrestrial setting. It is hard to estimate the total volume for the Pleistocene dune rocks on Isla del Carmen. To begin with, coastal erosion destroyed an unknown amount of the outcrop. While the thickness of deposits exposed at the coast is known, it is more difficult to ascertain precise dimensions farther inland. Nonetheless, a crude estimate for the volume of the dune rocks on Isla del Carmen is put at 57 billion cubic meters (Johnson and Backus, 2009). Assuming even a conservative estimate for limestone purity, the level of enrichment of the dune rocks calculated in numbers of articulated clamshells would be stupendous. It is certain that all the carbonate material held in the dune rocks had an organic origin with no other source than the clam flats situated directly north of the island.

Fig. 6.12. Pocket beach and dunes enclosed by Pleistocene dune rocks on Isla del Carmen.

Fig. 6.13. Fossil terrestrial snail from the dune rocks of Isla del Carmen (scale = 3 cm).

Pleistocene dunes buried the last coral reef that colonized the southwest end of Isla Cerallvo (Fig. 6.3, site Δ#3). Dune rocks composed of silica sand that retain cross-bedded structures migrated southward over the last of five consecutive cycles in reef development exposed in an 8.5-m thick coastal sequence on the island (Tierney and Johnson, 2012). Part of this sequence records steeply dipping leeward beds seen at shoulder level in the accompanying photo (Fig. 6.14). A coral sampled from below the dunes was submitted for laboratory testing and yielded an age of 122,000 years. Hence, the age of the ensuing dunes agrees with the timing of the last interglacial epoch. Like the nearby modern beach and sand dunes (site #12 described above), the source of silica sand comes from stream erosion of the island's mostly granite terrain.

Pliocene relationships. Carbonate dunes from the north side of Punta Chivato (Fig. 6.3, site □#1) are extensive and deeply placed within one of the promontory's larger canyons that penetrates from the Ensenada El Muerto south over a distance of 400 m. Fluvial action was undoubtedly the cause of the canyon's erosion, confined within steep walls mainly of andesite that rise to an elevation 80 m above sea level. Development of the canyon was likely to have accelerated during a period when sea level stood lower than today. An 11-m marine terrace that represents a higher stand in sea level than today cuts through dune rocks at the entrance to the canyon. This relationship indicates that dunes accrued sometime prior to the last interglacial ep-

Fig. 6.14. Burial of a Pleistocene coral reef by beach and dune deposits on Isla Cerralvo.

och. The most remote dune-rock deposit occupies the end of the canyon, where the landscape opens into a small interior basin. Winds that funneled into the valley apparently fanned out to follow the terrain, because a dune structure preserves the foreset layers that dip 28°to the southwest on a slip face (Russsell and Johnson, 2000). The terminal dune now sits at an elevation 40 m above sea level. This particular dune is perforated by many vertical hollows with a diameter of 5 to 10 cm that echo the shape and texture of small trees buried by the advancing dune.

On the west margin of the canyon, another body of dune rock occupies a position with a maximum elevation 60 m above sea level. Here, the vertical sidewall of the canyon is cut directly into the dune rock as opposed to andesite basement rock. The relationship shows that multiple dunes, some larger than others, filled the canyon over time and that part of the canyon's erosional history entailed re-excavations that eroded through dune rock. There is no trace of older Pleistocene terraces anywhere within the valley, which implies that dunes must have filled the valley during most if not all of Pleistocene time (Johnson and Backus, 2009). The inference of an older Pliocene age for the dune rocks is not certain, but possible based on the complex stages of development and overlap that occurred over time.

The dynamics of atmospheric circulation that push wind-driven waves from north to south over the full length of the Gulf of California with its full fetch is manifest in the development of sandy beaches and associated coastal dunes. In particular, the strong winter winds that funnel down the axis of the gulf have an impact on beach deflation that transfers sediment to adjacent dunes. The southward migration of nearly all the many dunes studied in the Gulf of California attests to the virility of the winter winds. While fossil dunes are not known to be as extensive as other kinds of geological deposits, sufficient evidence exists to demonstrate that high-pressure cells and attendant marine circulation maintained a strong influence over the gulf for millions of years.

Chapter 7

Mangrove Ecosystem (Open Lagoons)

Introduction

Small trees and shrubs that take on the role of classic umbrella species are at the core of the mangrove ecosystem in a transitional setting between fully terrestrial and marine systems. The mangrove forest is most widespread on coastal plains in the humid tropics near the mouths of rivers or around islands, coastal lagoons and estuaries. A marked connectivity exists among mangrove, sea grass, and coral reef ecosystems in tropical regions throughout the world. Generally, the growth of mangrove vegetation is limited to such climates where the average monthly minimum air temperature is ≥ 20° C (≥ 68° F). As in the case for coral reefs (see Chapter 3), the winter 20° C isotherm of adjacent seawater probably plays an important part in prohibiting the expansion of the mangrove ecosystem to higher latitudes. Within the Gulf of California, scattered colonies of the mangrove forest have adapted for survival in a far more arid setting and they represent the northernmost populations established on the Pacific coast of Mexico (Riosmena-Rodríguez et al., 2015). Like mangrove communities found elsewhere, the dominant vegetation is a key contributor to a carbon sink that transforms chemical and biological materials and provides a safe nursery environment for important fish stocks and other marine organisms.

The fossil record for mangroves is rich, consisting mostly of microfossils (pollen grains), but also macrofossils (wood, leaves, flowers, and fruit). Fossils from the Late Cretaceous Period suggest that mangrove species (mainly in the genus *Nypa*) originated along the shores of the Tethys seaway widely spaced between present-day Borneo, Bangladesh, Nigeria, and Venezuela (Ellison et al., 1999). Many more genera, including those belonging to the extant red and black mangroves, became widespread by Late Eocene time 34 million years ago. These several fossil taxa are thought to have occurred in the same sort of tropical, coastal swamps that now support extensive mangrove forests. The fossil record for the Gulf of California shows that mangrove plants migrated to the region not long after the major tectonic separation of the Baja California peninsula from mainland Mexico in middle Pliocene time about 3 million years ago.

Mangrove Zonation

Coastal zonation of the mangrove forest is affected by tidal range, land elevation, as well as soil and water salinity. Red, black, and white mangrove trees (*Rhizophora mangle*, *Avicennia germinans*, and *Laguncularia racemosa*, respectively) may all be found separately or growing in close proximity on the same shoreline. When found growing together as for example at El Requesón in Bahía Concepción (Fig. 7.1), each species is limited to a separate narrow zone defined by distance and changing conditions inland from the shore. Red mangroves occur at the water's edge, with full exposure to tidal variations and wind-driven waves. They are well adapted to these conditions with **prop roots** that extend outward and downward from the trunk to branch out like tripods. The tangled root system serves to increase stability as well as to facilitate the capture of sediments brought by waves and currents. The common name for this species comes from the brownish-red color of the smooth bark on the prop roots. This species also produces a pink flower. Moving further inland, the next zone is occupied by the black mangrove, with characteristic **pneumatophores** that stretch upward from the soil surrounding the trunk. These root adaptations function to supply oxygen to the underground roots that typically are buried in anaerobic (oxygen-free) sediment. Thousands of such aerated rootlets may service a single black mangrove tree. The common

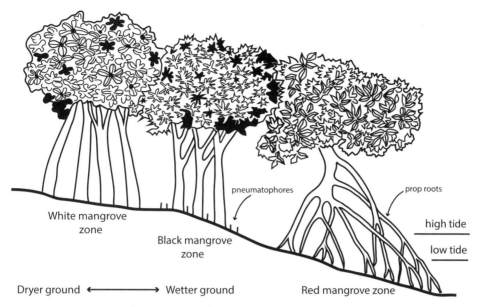

Fig. 7.1. Coastal zonation of mangrove vegetation (El Requesón).

name for this species is in reference to the color of the bark and underlying heartwood. White mangroves are favored by dryer soils (Fig. 7.1) and tend to dominate the interior of a mangrove forest. Small, bell-shaped white flowers give this species its common name. Pneumatophores may develop where conditions make the soil less aerated, but are lacking more often than not.

Plant Physiology

Bays with sizable tidal flats exposed under an arid climate typically register salinity levels far in excess of the 35 parts per thousand normal for seawater. Soils along such coasts may exhibit salinities in the range of 60 to 65 parts per thousand. For most plants, any salt content in the soil prohibits the uptake of water by roots and interferes with cellular metabolism. Mangroves have evolved a range of morphological and physiological traits that allow them to cope with such a harsh environment where other vascular plants fail to germinate. Elimination of salt and the demands for gas exchange that permit root function in an anaerobic setting are key to the success of mangrove plants (Medina, 1999). Water uptake from an environment enriched in highly permeable ions, mainly chloride and sodium, requires the capability for plants to remove salt ions at the root level or at the leaf level with compartments where salt can be accumulated with-

out damaging the photosynthetic apparatus. Open passageways (**aerenchyma**) found in the prop roots of red mangroves are essential for diffusion of oxygen. Transpiration in red mangrove leaves builds negative pressure in the vascular network of tubing responsible for circulation of sap. This activity induces a kind of reverse osmosis at the surface of the roots. Red mangroves also are able to store and dispose of excess salt in the leaves and fruit.

In black mangroves, oxygen diffusion is regulated by the vertical pneumatophores (interconnected by horizontal runners) that extend up to 20 cm above the ground surface as naked stems. These exposed peg roots have small openings (**lenticels**) that allow entry of oxygen, which is circulated deeper into the underground root system by the aerenchyma tissues. Black and white mangroves also excrete salt through specialized glands located on the surface of leaves. The ovate-shaped leaves of the black mangrove sometimes take on a silvery glint due to fine salt crystals that appear on the upper surface. Compartmentalization is crucial for separating salt-sensitive enzymes in the cytoplasm and cell organelles from the incoming salts in the transpiratory stream (Medina, 1999), and intracellular spaces called vacuoles are responsible for salt storage. When a plant has acquired too much salt by this method, elimination is achieved by dropping the leaves.

69

Propagation and Shore Stabilization

Contrary to other flowering plants with seeds that remain dormant for a period of time, the reproductive strategies of mangrove plants feature production of live seedlings via **propagules** that hang from the trees until they drop into water and float away. In the red mangrove, the propagules are large and cigar shaped (up to 15 cm in length). They may be seen dangling above the water almost any time of the year. Once dropped, a propagule floats vertically in the water and continues embryonic growth for as long as 40 days. In the black mangrove, the propagule takes on an elliptical shape like the seedpod of a lima bean, but smaller (2 to 3 cm in length). Production of propagules occurs only during the spring and summer seasons. Like the red mangrove, the black mangrove propagule floats in a vertical orientation, but embryonic development is restricted to a shorter period of time lasting only two weeks. The white mangrove has the smallest propagules (less than 0.5 cm in length) and the shortest period of embryonic development limited to about 8 days in seawater. Like the black mangrove, production of propagules occurs only during the spring and summer months. Dispersal in all cases is by waves and currents in seawater and the different kinds of mangrove propagules take root only after arrival in a suitable place such as a sand bar that is exposed from time to time by the tides. Thus, the mangrove forest sends out wave-borne seedlings that seek to expand the territory occupied and stabilized by these shore-dwelling plants.

Coastal wetlands dominated by mangroves influence physical processes on shorelines both indirectly and directly (Geadan et al., 2010). The exposed, above ground parts of the mangrove forest dampen waves where in direct contact with seawater and waterborne sediments. Plant stems and leaves slow water velocity, reduce turbulence, and promote deposition. The below ground contribution of decaying leaves and roots enriches soil organic matter. Fine, organic-rich soils tend to erode more slowly than mineral soils in wetlands (Gedan et al., 2010). Indirectly, mangrove forests retard coastal erosion by trapping and stabilizing shore sediments. Importantly, these wetland plants mitigate wave energy by building peat deposits and altering coastal bathymetry, a primary control over wave energy.

Biodiversity of the Mangrove Community

In a survey on the biodiversity from the fringing mangrove forest on the inner side of the great sand barrier El Mogote located on the Ensenada de La Paz (Felix-Pico et al., 2015), it was found that 78 species of macroinvertebrates are sheltered in a setting that would not exist without the umbrella effect of the mangrove trees. Among the many invertebrates, the black ark (*Anadara tuberculosa*) and the palmate oyster (*Saccostrea palmula*) register a significant presence due to high productivity and their potential commercial value. In particular, the oyster species at this locality shows an especially high level of abundance measured as 510 individuals per square meter. Oysters commonly attach to the prop roots of the red mangrove in large clusters, and are exposed during low tides as illustrated by an example from the leeward side of Isla Requesón in Bahía Concepción (Fig. 7.2). A detailed census on fish populations from the large mangrove colony at the southwest end of Isla San José (González-Acosta et al., 2015) found that 53 species belonging to 40 genera are affiliated with the ecosystem at that locality. Nearly 70% of those fish species have com-

Fig. 7.2. Oysters on prop roots of the red mangrove (El Requesón).

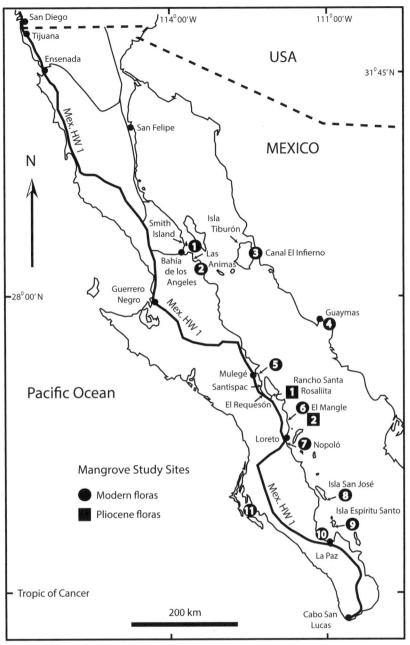

Fig. 7.3.   Study sites for present-day mangrove colonies in the Gulf of California including geological examples.

mercial value.   The importance of the mangrove ecosystem as a critical nursery for the region's juvenile fish stock is well demonstrated by this kind of study.

Gulf of California Mangrove Colonies

Living mangrove colonies within the Gulf of California occur in a scattered distribution along both shores with the size of individual colonies smaller, for example, compared to the extensive mangrove forest along the Caribbean shores of the Yucatan peninsula. Designated by numbered circles on the map in Figure 7.3, eight spots with notable mangrove colonies are marked along the east coast of the Baja California peninsula and its associated islands, two key localities appear on the opposite mainland shore of Mexico, and in the interest of completeness one locality is shown for the west coast of the Baja California peninsula.   Among these, the most northern documented mangrove colony on the Baja California peninsula is found

Fig. 7.4. Red mangroves lining the shores of the Mulegé estuary.

Fig. 7.5. Red mangroves at Santispac in Bahía Concepción.

on Smith Island (site #1), also known as Isla Coronado, in Bahía de los Angeles (Pacheco-Ruiz et al., 2006). There, a sparse stand of the red mangrove (*R. mangle*) with individual trees up to 2.5 m in height occurs on an inlet that penetrates deep into the island. A similar monospecific stand of the red mangrove also occurs along both sides of the inlet at Las Animas (site #2). Larger mangrove colonies are present along Canal El Infierno between Isla Ti-

burón and the Sonoran coast of Mexico (site #3), as well as the inner embayment at Guaymas (site #4).

Sheltered behind the local landmark known as Cerro El Sombrerito (site #5), the Mulegé River estuary and its mid-stream sand bar has long supported continuous stretches of the red mangrove (Fig. 7.4) but the setting suffered major damage due to flooding associated with the passage of tropical storm Juliette in 2013. It may take many years for the re-establishment of the mangrove colony. Farther south at Santispac within the partial shelter of Bahía Concepción (Fig. 7.3, site #6), a sizable colony features both red mangroves in a seaward position (Fig. 7.5) and black mangroves (*A. germinans*) in a more landward position. At the far end of the tombolo at El Requesón proximal to the leeward cliffs of Isla Requesón, (site # 5), it is possible to explore the full suite of red, black, and white mangroves depicted in Figure 7.1. The island is not accessible during high tide, but even from a distance it is possible to distinguish the outer zone red mangroves with their prop roots from the inner zone with white mangroves in the background (Fig. 7.6). White mangrove trees are fully developed on the north side of the tombolo close to the island cliffs (Fig. 7.7). At El Mangle (site #6), it is the white mangrove that entirely dominates the large arroyo mouth at the water's edge (see Plate 2). Driving south from Loreto to Nopoló on Mexican Highway 1 (site #7), the road passes alongside a golf course that impinges on the ultimate water trap with a prominent stand of red mangroves.

The southeast end of Isla San José (Fig. 7.3, site #8) features one of the largest patches of mangrove forest found anywhere inside the Gulf of California, covering an area of 121.8 ha or 300 acres (González-Acosta et al., 2015, p. 83). Several long coves that indent the western rocky coast of Isla Espíritu Santo set off pocket beaches that are partially colonized by red mangroves (site #9). Close to La Paz, the sheltered, landward side of the sand spit at El Mogote is dotted with patches of the red mangrove. On the Pacific Ocean coast, mangrove colonies are more extensive in the shelter of Bahía Magdalena (site #11). Commercial boat tours that

Fig. 7.6. Red mangroves (foreground) and white mangroves (background) at El Requesón.

Fig. 7.7. White mangrove trees at El Requesón.

take eco-tourists out to see the whales often include popular cruises among the mangrove thickets for birding.

Fossil Examples and Paleoecology

The extensive, worldwide fossil record for mangroves is based largely on recovery of hearty pollen grains extracted from muddy sediments. The outer walls of pollen grains are composed of a tough biopolymer substance highly resistant to decay. Occurrence of macrofossils such as woody stems, leaves, and fruits is rare, but the potential for preservation is much enhanced by chemical activity around mineral springs on the Baja California peninsula (see Chapter 10). Delicate woody structures of plants are subject to petrification through a slow, cell-by-cell replacement of woody fiber with opalline silica or other minerals circulated in fluids

from geothermal springs. So long as hydrothermal springs are located in a coastal area, such as found today at Santispac in Bahía Concepción, the likelihood is high for woody debris from mangrove plants to become fossilized in that special kind of setting.

Pleistocene relationships. Currently, no examples of fossil pollen characteristic of mangrove species are reported from Baja California. However, the possibility for making such a discovery is good and represesents a new line of research. A potential locality for investigation is the Pleistocene terrace that sits 12 m above the estuary of the Mulegé River with a recent history of a living red mangrove colony in place along the river bank below. Uplifted terrace sediments at this spot consist of poorly indurated, massive to poorly bedded silt with a high content of arkosic silica that includes coral fragments dated as 124,000 years old consistent with the last interglacial epoch (Ashby et al., 1987).

Pliocene relationships. A fossil locality with silicified remains of rootlets from the black mangrove (*A. germinans*) occurs within a well-bedded, 15-m thick chert deposit belonging to the Infierno Formation at Rancho Santa Rosaliita near the southeast corner of Bahía Concepción (Fig. 7.3, site □#1). In the context of a stratigraphic sequence with overlying limestone and underlying terrestrial deposits containing conglomerate, the chert can be understood as having formed in a cluster of shallow basins with a terminal water depth of 10 m or less (Ledesma-Vázquez et al., 1997). This interpretation

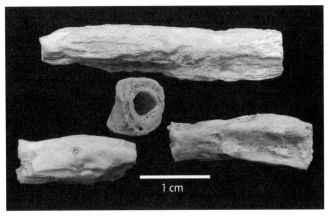

1 cm

Fig. 7.8. Fossil root fragments from a black mangrove in a chert deposit at Rancho Rosaliita.

73

is supported by fossilized segments of mangrove rhizomes (Fig. 7.8), the modified plat stem typically found underground as horizontal runners. In this case, the runners are preserved in growth position in the chert that can be traced continuously for a distance of about 50 cm. The stratigraphic sequence indicates that lowlands on the margin of Bahía Concepción were flooded by a rise in sea level during middle Pliocene time in agreement with the accepted scenario for mangrove colonization in coastal plains around estuaries and marine lagoons.

Appropriately enough, petrified woody debris is abundant at El Mangle north of Loreto (Fig. 7.3, site □#2). Along a paleoshore at the margin of the Cerro Mencenares volcanic complex, a distinctive deposit up to 2 m in thickness is exposed in rocks composed mainly of a friable mixture of silicified clay-size to sand-size particles that include conspicuous clots of opal (Johnson et al., 2003). Thin lenses of solid, gray-white opal occur throughout the deposit. The friable material yields abundant silicified plant twigs that account for as much as 20% of a given sediment sample. Most of the twigs, which are preserved well enough to retain the texture of their cortex cells, bear a striking resemblance to living *Atriplex* salt bushes growing at the mouth of Arroyo El Mangle. Less common are larger pieces of fossil wood that show tree limbs with recognizable knots and growth rings in cross section (Fig. 7.9). The largest trees still growing in the area are represented by a grove of the white mangrove (*L. racemosa)* that gives El Mangle its place name. It is logical that the larger pieces of fossil wood from this locality may belong to this species. A Pliocene age for this deposit is based on fossil echinoids (*Clypeaster marquerensis*) from the overlying limestone, but more precisely in the context of a tuff layer dated radiometrically as 3.3 million years old. The date puts the deposit squarely in the Piacenzian Stage of the Upper Pliocene. Similar

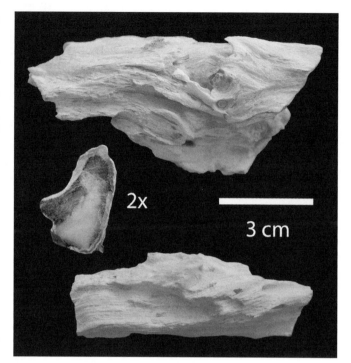

Fig. 7.9. Pliocene wood likely from a white mangrove replaced with opal at El Mangle.

to the chert deposit at Rancho Santa Rosaliita, the opal from El Mangle has a source related to a Pliocene geothermal spring.

The mangrove ecosystem within the Gulf of California is spotty in distribution and subject to local damage from hurricanes, but overall it is an integral part of nature that has a profound effect on the health of marine communities including a wide array of both invertebrate and vertebrate organisms. Without the valuable service of its nurseries established in physically sheltered spots, the greater biodiversity of the gulf would be diminished. For the paleoecologist, it is thrilling to know that the mangrove ecosystem gained a toehold on the gulf early in its tectonic development. There is ample opportunity for future studies that focus on microfossil evidence for mangrove colonization based on pollen profiles, particularly during the more recent history of Pleistocene times.

Chapter 8

A World of Microbes (Closed Lagoons)

Introduction

Among the least explored ecosystems in the Gulf of California is that in which diverse microbe communities are mostly confined to closed lagoons. Microbes (another word for bacteria) can withstand extreme changes in temperature, salinity, and desiccation living on the margins of such lagoons, where other marine creatures would find it impossible to survive should they somehow gain entry. On several islands and at other places along peninsular shores, waves and currents build strong berms from cobbles and boulders that physically isolate standing bodies of water from circulation with normal seawater. Based on a preliminary reconnaissance (Johnson et al., 2012), there are more than a dozen closed lagoons within the gulf region easily spotted from satellite images. Many other examples await discovery through a better-organized ground search. Under an arid climate with strong sun and high rates of evaporation, lagoon water is subject to rising levels of salinity. Some seawater is filtered into the lagoons through the enclosing berms. Influx of fresh water arrives only with the runoff of infrequent rainwater. A similar but artificial setting beneficial to the growth of microbes occurs in commercial salt ponds.

Easily overlooked compared to multicellular forms of life expressed by plants and animals, microbial communities typically form sticky mats or raised domes that trap and bind coatings of fine sediment. Because the individual bacterium is so small, the colonies within which microbes thrive are not very noticeable aside from the so-called organo-sedimentary structures they accrete. Dome-shaped structures with an outer jacket of living microbes were first recognized and studied in Western Australia at Hamlin Pool on Shark Bay along the Indian Ocean (Logan, 1961). The discovery was sensational, especially for geologists and paleontologists, because comparable structures are well known from the rock record tracing far back in geologic time. Now referred to as microbialites by biologists, paleontologists had known the same features as stromatolites. The latter term comes from Greek roots, meaning "stony blanket." The oldest such structures date from Archean time 3.5 billion years ago, when the atmosphere of planet Earth was unlike ours today. The big difference was a complete lack of oxygen. At that time, stromatolites were not limited to places where water salinity was unusually high. From research on a rock formation in Western Australia called the Strelley Pool Chert (Allwood et al, 2006), it is understood that stromatolites were adapted to a range of physical settings including rocky shores, lagoons, reefs, and open seafloors.

In particular, cyanobacteria make a living by photosynthesis, the byproduct of which is oxygen. As stromatolites were once very widespread, paleontologists believe they played a leading role in altering the Earth's atmosphere through the gradual build-up of oxygen. Living cyanobacteria may not be as scarce as once thought, and continue to play some role as atmospheric biofilters (Margulis and Sagan, 1987). Mat-forming microbialites were first described from a lagoon on the west coast of Baja California during the mid-1970s (Horodyski et al., 1977). Similar living mats were found much later in the Gulf of California (Johnson et al., 2012), where they appear to be more widely dispersed than on Mexico's Pacific shores.

Microbes in the Tree of Life

In our minds, the world appears dominated by two biological kingdoms: that of animals and that of plants. In fact, biologists consider that life is divided into several kingdoms with three more generally characterized by fungi, by protists, and by bacteria (Fig. 8.1). Whether large or small, plants and animals share several qualities in common. They consist of many kinds of cells, all of which feature a nucleus enclosed by a separate membrane within the greater cell. Cell division occurs by mitosis, the process by which chromosomes in the nucleus become separated into two identical sets, each with its own cell nucleus. There is sexual differentiation (male and female gametes). Animal and plant cells perish in the absence of oxygen. The same traits apply to the fungi, which include members of the familiar mushroom family as well as yeasts and molds. The protists comprise a fourth kingdom, considered problematic by some biologists. Members are mainly unicellular in organization, but also include multicellular representatives. They live exclusively in water but otherwise share all the same traits of cellular organization enjoyed by animals, plants, and fungi.

In terms of overall diversity, the kingdom of the bacteria is by far the largest although much remains to be accomplished in the study of microbe species. Bacterial life, which prominently includes the cyanobacteria, is exclusively unicellular. Compared to life forms in the animal, plant, fungi, and protist kingdoms, the nucleus of bacterial cells is loosely contained within the cell and not bound by its own membrane. Cell division is by binary fission, which effectively produces identical clones. This level of cellular organization has its own designation as a prokaryote form of life, distinct from the other kingdoms that are characterized as eukaryote life forms. In reference to the bound nucleus common to all animal, plant, fungi, and protist cells, the word eukaryote derives from Greek roots, meaning "well + kernel" as in a shelled nut. The term prokaryote also derives from Greek roots, meaning "before + kernel" in reference to this most ancient stem group in the tree of life. The archaea is yet another class of microbe-like organisms, now regarded as a distinct domain. An important difference is that bacteria live in oxygenated environments, whereas the archaea are anaerobic (Fig. 8.1). Like bacteria in some ways, the archaea are unicel-

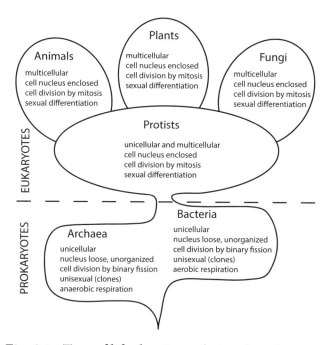

Fig. 8.1. Tree of life showing relationships between prokaryote and eukaryote groups.

lular prokaryotes that reproduce by asexual means. In terms of geologic time, the archaea predate the bacteria, although both kinds of prokaryotes were capable of forming organo-sedimentary structures. Stromatolites reached a peak in global domination more than one billion years ago, but went into steep decline as multicellular plants and animals appeared in greater diversity starting about a half-billion years ago. Cyanobacteria continued to leave a fossil record across geologic time, including examples from Baja California. A related group that co-existed with marine invertebrates long after the great falloff in stromatolites is called thrombolites. These differ from stromatolites in two ways. First, they lack the fine-scale layering or laminations typical of stromatolites. Instead, they exhibit a lumpy or clotted fabric wherein any trace of layering is indistinct. In fact, the name is derived from the Greek word thrombos, meaning lump or clot. Secondly, this kind of bacteria is able to form a calcium-carbonate crust that thickens over time with growth.

Stromatolite Form and Function

Fossil stromatolites seldom reveal any trace of the original microbes that contributed to the development of organo-sedimentary structures. Thus, living cyanobacteria from places like Shark Bay (Australia) or the closed lagoons in the Gulf of California provide critical insight regarding the fundamentals of stromatolite design and function typical through much of geologic time. Fine-scale layering in stromatolites is characteristic, whether expressed in the shape of a dome, a pillar, or a flat mat. Domes are common in the near-shore, shallow waters of Shark Bay's Hamlin Pool, where waves and tidal currents affect variations on that theme. In some cases, the shape is less circular in plan view and more elongated like a loaf of bread. This is the result of strong tides that sweep back and forth perpendicular to the shoreline. Although wind-blown waves may be influential in larger lagoons, tidal action is limited in the stromatolitic lagoons from the Pacific shores of Baja California. Moreover, tidal action is entirely absent in the closed lagoons from the Gulf of California. As a result, the shape adapted by microbialites in closed lagoons is that of a planar mat.

Where development within a protected lagoon has continued for a long time, the mats follow one after the other like a pile of carpets. To carry the carpet analogy farther, the mats are composed of filamentous cyanobacteria that produce tread-like strands similar to the tufts that stand up on a carpet. The filaments act as a mucus-coated sheath that encases a string of cells (Fig. 8.2). Individual strands grow in length as neighbor cells encased within the sheath expand through binary fission. While individual strands may become quite long, filament diameter is on the order of only 10 μm. Once the strand becomes too long, breakage tends to occur at vacant places between cell clusters. Cells inside the filaments use sunlight in the process of photosynthesis, but sunshine carries high doses of ultra-violet light harmful to cells. It is believed that the mucus sheath functions to protect the microbes from UV exposure. An unrelated consequence is that sticky filaments trap and bind together fine particles of silt and clay that enter the lagoon as wind-born sediment. Individual filaments may immerge from beneath a light coating of sediment to continue the task of photosynthesis. Long after a carpet of microbes has expired and its organic

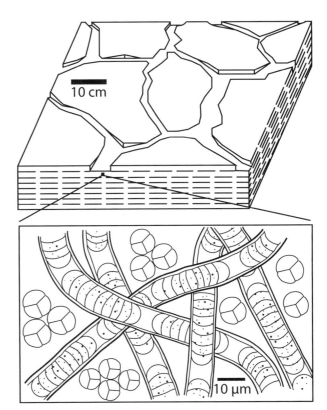

Fig. 8.2. Schematic showing microbial mats with desiccation cracks (above) and detail from layers formed by filamentous and coccoidal bacteria (below).

matter has lost any trace of cellular organization, a fine succession of sediment layers remains in place to record the accretion of an organo-sedimentary structure.

In addition to the strands of cells linked together within a filamentous sheath, another group of cyanobacteria occurs as ovoid or spherical-shaped cells (called a coccus) individually as much as 10 μm in diameter (Fig.8.2). The cocci (plural) may cluster into larger aggregates, but are never contained in a shared sheath or bag. Such coccoidal bacteria commonly coexist with filamentous bacteria in a diverse community. They represent examples of extremophiles able to withstand extreme variations in temperature, high salinity, and the ionizing effect of high dosages of UV light. In situations where a mat becomes excessively thick, photosynthesis is no longer a viable option for the more deeply buried microbes. In that case, members belonging to the archaea domain may proliferate in a hidden anaerobic state.

Microbial Mats in Baja California

The accompanying map of Baja California (Fig. 8.3) marks several localities (circled numbers 1 to 8) where living microbial mats have been documented in the published literature or are known to exist. On the Pacific coast in northern Baja California (site #1), the locality first described by Horodyski et al. (1977) at Laguna Mormona is included for the sake of completeness. An extensive salt flat borders the closed lagoon on the landward side and barrier sand dunes close off the lagoon on the seaward side. Laminar mats accreted by filamentous cyanobacteria are described as present (*Microcoleus chthonoplastes* and *Lyngbya* sp.). At least one species of coccoidal bacteria (*Entophysalis* sp.) also is present. Mats grow in water up to 30 cm deep, but are subject to tides and other fluctuations that make them vulnerable to desiccation. Microbial mats become dry and cracked under the sun, where large parts

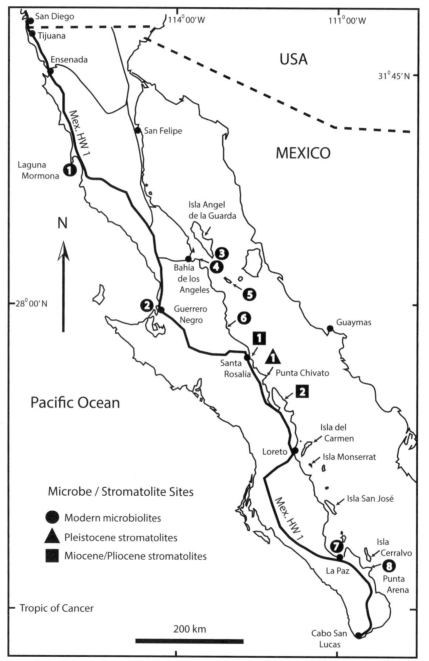

Fig. 8.3. Map showing reference sites for living microbialites and fossil stromatolites in Baja California.

of the lagoon are exposed around the edge of the salt flats. Desiccation polygons typically measure a half-meter in diameter (Horodyski et al. 1977), much as depicted in Fig. 8.2.

The tidal flats at Guerrero Negro (site #2) also have received considerable attention with respect to the occurrence carpet-like microbialites. Similar to Laguna Mormona, the Guerrero Negro mats are dominated by filamentous cyanobacteria (*Lyngbya* sp.), but another kind of cyanobacteria (*Calothrix* sp.) also forms mats with a pustular surface closer to shore. The microbial mats from both Pacific coast locations exhibit no evidence of mineralization, and therefore show no potential for fossilization. Studies on the Guerrero Negro microbialites are more focused on biochemistry (Des Marais, 2003).

Several closed lagoons occur around the periphery of Isla Angle de la Guarda in the northern Gulf of California. Two at the southeast end of the island feature extensive microbialite constructions that represent the first living thrombolite and stromatolite associations identified in the region (Johnson et al., 2012). Separated by only 1.5 km, they are labeled as parts of the same locality on the accompanying map (Fig. 8.2, site #3). The larger lagoon has an oblong layout that extends for a distance of 1.25 km. Informally called the Big Lagoon, it is the largest such closed lagoon detected anywhere on the east side of the Baja California peninsula. Plate 3 shows a view to the north across the pale-blue water of the lagoon and long outer berm with darker blue water in the open gulf beyond. Dark patches of soil free of vegetation appear in the foreground. These sit over depressions in the topography and probably represent a series of pre-existing lagoons that became elevated during local uplift of the island. The Big Lagoon provides a useful model for all such closed lagoons with extensive microbial development in the Gulf of California (Fig. 8.4). In this model, mineralized thrombolites form an outer ring around the lagoon with laminar stromatolites occupying an inner ring. Soft stromatolites are concealed beneath a thin carbonate crust deeply scored by desiccation cracks. Generally, the com-

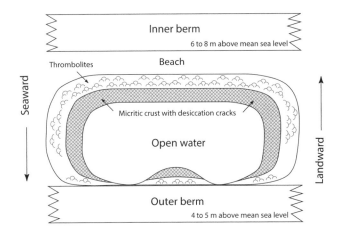

Fig. 8.4. Model for thromboblites and microbialites from closed lagoons (after Johnson et al., 2012).

bined width of the two bands is from 4 to 6 m wide around the margins of a lagoon with open water in the center.

At the Big Lagoon, the outer protective berm is massive with a 50-m wide base that sits from 4 to 5 m above mean sea level (Fig. 8.5). Loose gravel on the berm's inner slope shows signs of disturbance by waves that over-topped the structure perhaps during the confluence of the last storm during a high tide. Such marks indicate that the lagoon is refreshed with normal seawater from time to time. Also, the gravel is porous enough that some seawater seeps into the lagoon through the berm's base. An incised out-flow channel is located at the east end of the Big Lagoon, like a spillway still below the lowest part of the enclosing berm. This feature indicates that the lagoon has a history of flooding

Fig. 8.5. Outer protective berm on the Big Lagoon (Isla Angel de la Guarda).

from rainwater runoff during major storms that strike the area from time to time. Hence, the degree of salinity in lagoon water is assumed to fluctuate somewhat but mostly remain hypersaline over long intervals of time. The salinity of the nearby Small Lagoon was recorded as 148 ppt in November 2007 (Johnson et al., 2012).

Mineralized thrombolites from the outer ring at the Little Lagoon on Isla Angel de al Guarda have a knobby surface with a maximum relief between 2 and 3 cm that superficially looks like small cauliflower heads stripped of all leaves and compressed into a single compact sheet (Fig. 8.6). As in the Little Lagoon, a thin crust with mineralization that is deeply scored by desiccation cracks overlies the soft laminar layers of stromatolites on the Big Lagoon (Fig. 8.7). Biological samples cultured from soft laminar stromatolites from the Little Lagoon's inner ring were found to include abundant examples of the solitary, coccoidal cyanobacteria (*Chroococcidiopsis* sp.) together with several filamentous cyanobacteria (*Phormidium* sp., *Oscillatoria* sp., *Geitlerinema* sp., *Chroococcus*, and perhaps *Spirulina* sp.). Examples showing strands of filamentous bacteria (*Phormidium* sp.) and a cluster of the coccoidal bacteria (*Chroococcidiopsis* sp.) are illustrated in Plate 3. The same community of cyanobacteria was found to inhabit the closed lagoon at Ensenada El Quemado (site #4) on the peninsular mainland south of Bahía de los Angeles (Johnson et al., 2012).

Another closed lagoon with microbialites is located near the southwest end of Isla San Lorenzo (Fig. 8.3, site #5). The passage between San Lorenzo and the peninsular mainland is the Canal de Saispuedes, known for its strong tidal currents. Shown in Plate 3, the aerial view over the southern part of Isla San Lorenzo captures the scale of the lagoon against the island's rugged topography. The berm that isolates the lagoon from the adjacent channel was formed by gravel carried southward along the coast through the action of a vigorous long-shore current. The lagoon has an open surface area of 22,000 m² surrounded by a rim with a distinctly mineralized crust scored by desicca-

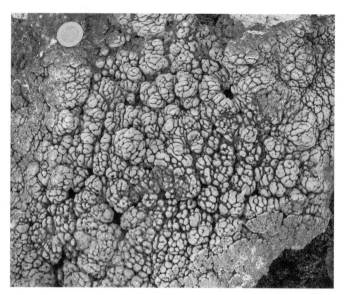

Fig. 8.6. Detail of thrombolites from the Little Lagoon (Isla Angel de la Guarda). Coin for scale is 2.4 cm in diameter.

Fig. 8.7. Floating microbialites scored by desiccation cracks (Isla Angel de la Guarda).

tion marks (Thomas Bowen, personal communication, June 2014). The scenario is comparable to that found in the similar-size lagoon at Ensenada El Quemado (site #4) to the north.

Situated on the gulf coast 12 km south of the border between Baja California and Baja California Sur, Bahía San Carlos is an interesting spot (Fig. 8.3, site #6), where salt ponds look to have been excavated as part of an extensive commercial operation. The closed lagoon at the seaside below the salt ponds is easy to detect on an over-flight, due to the reddish-brown coloration of the lagoon. Although yet to be confirmed, it appears likely that extensive microbial colonization has occurred at this locality.

Far south at La Paz (Fig. 8.3, site #7), the city's harbor area is sheltered by an 11-km long, club-shaped sand spit called El Mogote. This enormous geomorphic feature terminates with an eastward-directed peninsula that approaches to within one kilometer of closing off the Ensenada La Paz. Pro-thrombolite structures formed by cyanobacteria are identified at several localities abound the circumference of this restricted bay (Siqueiros-Beltrones, 2008). The structures are regarded as "pro-thromobitic" because they lack the fine-scale laminations found in microbial mats, but are not mineralized like thrombolites. They are described as clotted, semiconsolidated biosedimentary structures attributed to the binding action of filamentous cyanobacteria with mucus-coated sheaths (*Microcoleus chtomoplastes*, *Oscillatoria limosa*, and *Lyngbya aestuarii*). It is worth noting that microbialites are well developed in one of the most sheltered spots at Estero Zacatecas in the northwest corner of the lagoon, where a thicket of mangrove trees stabilizes the narrow part of El Mogote.

Modern microbialites are sure to be found many places around the Gulf of California, but the final example in this section comes from the commercial salt ponds at Punta Arena de la Ventana (Fig. 8.3, site #8) situated 15 km east of the town of La Ventana. There, it is possible to explore a grid-work of shallow salt ponds, some of which have gone unattended for a long time. Some vacant ponds feature an uneven mat of reddish-brown organic material, in part with a pustular surface like a rumpled carpet (see Plate 3). The material is not mineralized in any way, but has some internal integrity and can be pealed away in sheets. On the whole, the coloring is comparable in tone and in

setting to microbial mats from Brazil that successfully were transferred to an aquarium for controlled study (Vasconcelos et al., 2014). Essentially, the artificial salt ponds at Punta Arena provide natural aquaria that have a greater capacity. The particular setting has much potential for future study of microbialites.

Fossil Examples and Paleoecology

Much is anticipated in future years through the on-going search for examples of stromatolites as fossil representatives of living microbialites in the Gulf of California. A good start already has been made with respect to one Pleistocene and two Pliocene examples, registered on the accompanying map (Fig. 8.3) by a numbered triangle and numbered squares, respectively.

Pleistocene relationships. Limestone laminations identified as belonging to a stromatolitic mat are associated with an uplifted marine terrace at Punta Chivato (site ∆#1). Prior to a breach that allowed the introduction of intertidal invertebrates (such as the gastropod *Turbo fluctuosus*), thin sheets of limestone ranging from 5 mm to 450 mm in thickness covered an area >1,000 m^2 within a closed depression on andesite bedrock (Backus and Johnson, 2014). The sheets may have been anchored to the margin of the pool, but probably floated over shallow water for the most part. This interpretation is based on the fact that the limestone crusts are not now continuous, but broken into slabs up to 30 cm x 20 cm in size strewn over the ground (Fig. 8.8). Nothing is found encrusted directly onto the rock floor.

This fits the scenario from the Big Lagoon on Isla Angel de la Guarda, where living microbial mats are attached to the shore but extend a short way out over the water as a floating mass not unlike bog vegetation. In cross section (Fig. 8.9), the crusts have the superficial appearance of plywood warped by immersion in water. Wavy laminations, some of which are separated by cavities, suggest that gas was generated as the microbes became degraded under conditions of natural burial. Observed

Fig. 8.8. Slabs with Pleistocene stromatolites (hammer, dog, car for scale) from Punta Chivato.

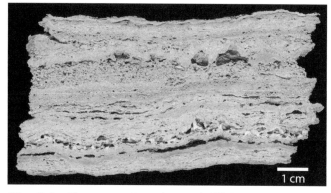

Fig. 8.9. Cross-section through Pleistocene stromatolites from Punta Chivato.

under magnification with a scanning electron microscope, specific evidence in the form of calcified filaments and high-magnesium microcrystalline calcite points to the connection with cyanobacteria (Backus and Johnson, 2014). Subsequent testing for biochemistry confirms that biomarkers for cyanobacteria and sulfur-reducing bacteria were detected in a sample under laboratory analysis (Jörn Peckmann, Vienna University, personal communication, February 2015). However, no traces of

biomarkers from the domain of the archaea were found, as might be expected from deeper within a microbialite deposit.

The remaining basin rim surrounding the fossil deposit now resides at an elevation 75 m above sea level near Punta Chivato. Exact dating of this particular deposit is imprecise, but the 12-m marine terrace cut in the rocky shore immediately below correlates to the last interglacial epoch about 125,000 years ago. It is estimated that the 75-m terrace with the laminated limestone corresponds to an interglacial epoch dating from 334,000 or even 712,000 years ago. The Pleistocene glacial and interglacial epochs are defined on a global time frame related to overall cycles in the Earth's fluctuating temperature as the continental glaciers in the northern hemisphere advanced and retreated through time. Glacial ice never came close to Baja California, but shifting patterns that brought a cooler and wetter climate farther southward did impact the Gulf of California on a recurring basis. Thus, discovery of the Pleistocene stromatolites from Punta Chivato has ramifications that relate both to climate change and to the ongoing tectonic forces shaping the Gulf of California.

Miocene to Pliocene relationships. Local conditions in the Santa Rosalía area after the start of the Pliocene promoted the mineralization of travertine or tufa deposits from seawater saturated with calcite. Earlier during the latest Miocene around Isla San Marcos and Santa Rosalía, extensive beds of gypsum were deposited due to evaporation of the earliest seawater to reach the protogulf in this region. Traces of late Miocene stromatolites associated with travertine are described from exposures in the sea cliffs at El Morro (site □#1) on the south side of Santa Rosalía (Miranda-Avilés et al., 2005). The stromatolites at this locality preceded arrival of marine invertebrates that denote normal levels of marine salinity.

A more extensive deposit of reputed Pliocene age occurs on the east side of the Concepción peninsula, where a closed lagoon with an area of 4 km² is situated 300 m south from the mouth of Arroyo El Coloradito (Fig. 8.3, site □#2). Described by Johnson and Ledesma-Vázquez (2001) as part of their survey between Punta Chivato and Punta San Antonio, the spot is reached by boat 26 km southeast of the northern tip of the Concepción peninsula. At the back of the lagoon, there is a 3-m thick deposit of gypsum crusts individually from 3 cm to 10 cm in thickness separated by laminations of very fine-grained limestone. The gypsum crusts are curved in cross-section with raised domes and merged margins up to 1.5 m in diameter (Fig. 8.10). In their original description of the site, the fine limestone partings between gypsum layers were explained as micrite deposits left by calcareous green algae. Desiccation cracks with a diameter of 0.5 m are imprinted on the exposed upper surface of the gypsum deposit.

Fig. 8.10. Gas domes with interlayered gypsum and microbialites near Arroyo El Coloradito on the east side of the Concepción peninsula.

Comparison with a study by Aref et al. (2014, their fig. 8) on structures from the salt works at Ras Shukeir on Egyptian shores of the Red Sea makes a strong case that the raised structures from the Concepción peninsular are gas domes featuring interlaminated gypsum and fine limestone left by microbial mats. It is inferred that repeated cycles of wetting and drying promote the formation of inter-bedded microbialites and gypsum, respectively. Domes arise when pressure from gas generated beneath sticky microbial mats fails to escape through the overlying layers and causes those layers to part from the substrate. The lifting of the domes leaves empty voids behind in the structure once it has solidified. Gas domes from the closed lagoon near Arroyo El Coloradito are larger that those reported from Egypt, but otherwise fit the description of the contemporary features very well.

A larger evaporitic lagoon than the present must have existed at the site where the gypsum domes are exposed in cliffs at an elevation about 9 m above sea level near Arroyo El Coloradito. Due to a lack of associated fossils, no firm date for the gypsum domes can be assigned, but incorporation within a ramp-like structure above the surrounding area argues in favor of a Pliocene origin (Johnson and Ledesma-Vázquez, 2001).

Discovery of contemporary microbialites that form extensive mats in closed lagoons and salt ponds throughout the Gulf of California and their linkage to precursors that date back several millions of years represents a window on the past regarding a fragile ecosystem. Because of its isolation from other ecosystems in the gulf region and the minimal balance of nutrients that circulate through a segregated world of microbes, the scenario is regarded as an oligotrophic ecosystem. Derived from Greek roots (oligos and trophikos = "few + feeding"), the term refers to an environment that offers little to sustain life. Given what yet remains to be learned about the distribution of the closed-lagoon ecosystem in the Gulf of California, several intriguing questions can be explored.

Do all closed lagoons in the gulf region support microbialites, and if not why not? How do the simple prokaryotes that comprise the cyanobacteria and other microbes manage to colonize any given closed lagoon or body of water with a limited opening to the sea? How do the microbes spread from one setting to another? What roles are played by short-term and longer-term variations in the region's climate in the viability of this ecosystem? In particular, what role might seasonal winds play in the circulation of tough germ cells from microbes that might aid in their air-borne transit? Such questions point the way ahead to a field of study attractive to exobiologists interested in exploring the potential for life beyond planet Earth.

Chapter 9

Estuaries and Delta Systems

Introduction

The Colorado River empties into the Gulf of California from an enormous drainage area in the southwest part of North America covering 630,000 km². Dams on the river have halted the growth of this once vibrant delta system (Carriquiry and Sánchez, 1999), which still occupies an area larger than 800 km². By comparison, the Baja California peninsula is not known for rivers of any size that flow throughout the year and contribute to the continuous buildup of shoal-water deltas. Emptying farther south at Mulegé, the spring-fed Mulegé River is one of the few continuously running streams to reach the gulf from the peninsula's eastern slope. The combined channels that flow to the Mulegé estuary define only a small drainage basin of 250 km² (Skudder et al., 2006). On the whole, it may appear that the lands around the Gulf of California offer a poor prospect for the study of estuaries and dynamic delta systems. On the contrary, arroyos trace a network of stream courses extensively spread throughout the Baja California landscape. However, the flow of water in those channels is ephemeral and the chief limitation to delta development is due to low rates of precipitation over most of the region. A more substantial pattern of rainfall during the geologic past is indicated by thick sandstone and conglomerate beds that occur at scattered localities between Isla Angel de la Guarda in the north and Isla Cerralvo in the south.

Thick sand and gravel accumulations with a particular configuration are key to the identification of delta deposits, whether formed under present-day conditions or during the remote past. Generally, the pebbles, cobbles, and small boulders that make such deposits are worn and rounded by physical abrasion during high-energy flood events that transport river sediments downstream. Other deposits with rounded cobbles and boulders might be confused with delta deposits, in particular those that accumulate along rocky shores under high-energy wave attack (see Chapter 2). However, rocky-shore conglomerate typically occurs in comparatively thin layers having a broad lateral distribution in direct contact with source rocks below. Fossils may be found on individual clasts in a rocky-shore deposit, due to the lifestyle of inter-tidal organisms like barnacles or oysters that attach to rocks bathed by normal seawater. In Baja California, deposits that follow after a rocky-shore conglomerate usually entail limestone-forming organisms such as corals, bivalves, and rhodoliths (see Chapters 3 to 5). In contrast, fossils seldom are found in river or delta deposits. Moreover, delta deposits are far more restricted in spatial distribution.

Delta Shape and Form

In map view, the shape of a typical river delta on meeting the open sea is that of a triangle or fan, with the fan handle fixed in the river's mouth, and the rest spread outward from that point in a radial pattern. The very term, delta, derives from the triangular-shaped Greek letter by that name. The word was commonly used by Greek geographers long ago to describe river deposits at the edge of the sea. The most famous of the early Greeks to apply the term was the historian Herodotus, who recorded a detailed description of the Nile River delta based on direct observation sometime about 440 B.C. In practice, the Nile delta offers a basic model that can be applied with few variations to most such deposits all around the world (Fig. 9.1). As a river flows seaward on a gradient, it tends to shift its main channel from season to season within the confines of a flood plain. On nearing the sea, however, the river's flow may diverge into a few or several **distributary channels** that fan-out beyond the river mouth. In part, this is due to a decrease

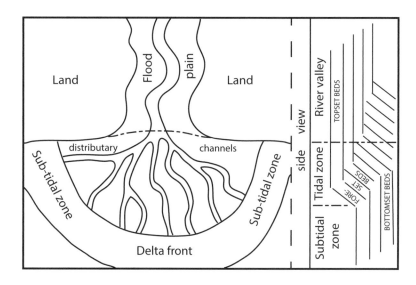

Fig. 9.1. Fan-delta model in map view and cut-away side view with prograding topset, foreset and bottom set deposits.

in stream velocity as river water flows into a much larger but relatively still body of water. Diversions also are influenced by the semi-circular shape of the **delta front** as it pushes outward into deeper water. Side channels may develop in order to take advantage of a shorter path to the sea.

      In side view (Fig. 9.1), layering within a delta deposit exhibits three types of beds. **Topset beds** are deposited as horizontal layers on and around the flood plain and inner parts of the distributary channels. **Foreset beds** accumulate on an incline across the distal parts of the distributary channels and the delta front. **Bottomset beds** occur as horizontal layers in deeper water out beyond the delta front. As an active delta continues to add new materials from upsteam and builds farther seaward, flat-lying topset beds may overtake and truncate older foreset beds. Likewise, the addition of new foreset beds may overlap older bottomset beds (Fig. 9.1). The fundamental form taken by a fan delta is that of a sedimentary bulge with an arched profile, as viewed laterally in cross-section from one side of the delta to the other. Such a profile is bowed in shape, because topset beds build upward on the central part of the delta at the same time that foreset beds build outward around the flanks. A delta associated with a great river, like the Amazon River in South America, can be expected to be very large. Even so, all deltas are limited in size by the amount of material carried seaward and other factors in the context of regional climate and physical geography. The more rainfall, the greater the rate of erosion. The larger the drainage basin, the greater the amount of sediment susceptible to erosion.

Delta Construction Variables

An adequate sediment source, the gradient of the river course emptying into the ocean, and the degree to which the river is channelized as opposed to being unconfined account for the key variables in delta construction (Postma, 1995). Four major delta types fall into a classification scheme based on differences in fluvial style. Immature streams on a very steep gradient with little or no channelization produce mass flow of coarse sediments leading to a type-A delta. Streams on a somewhat lower gradient (4° slope) with an unstable bed load of coarse sediments lead to a type-B delta. Steams characterized by a moderate gradient and relatively stable bed load of finer sediments empty into a type-C delta. Mature streams with a very low gradient and a stable load of fine sediment carried in suspension typically are constrained by levees and lead to a type-D delta. A mature stream might be rejuvenated through tectonic uplift of the surrounding uplands. As a consequence, a type-D delta may be transformed into a type-B or even a type-A delta. Under conditions of episodic uplift along a coastal zone, conditions conducive of a type-B delta may be perpetuated over a long time. Delta construction also may be influenced by the condition of the lower **shoreface** along the coastal zone, representing a place well below the low-tide mark where sediment is less subject to disturbance by waves. Whether or not the shoreface plunges into deeper water or descends gradually on a lesser incline makes a difference in the form of a growing delta system at that intersection.

Climate variation is the other principal factor that has a profound effect on delta construction. A delta system is starved of sediments under an arid climate, typically in place around 30° north or south of the equator. Deltas that occur in tropical or temperate rain zones are insured a steady source of sediments. The northern part of the Baja California peninsula and adjacent Gulf of California sit squarely in an arid zone, whereas the southern Cape Region and the opening of the Gulf of California on the Pacific Ocean belong to a zone that is semi-tropical in nature. The region is impacted by a seasonal, semi-monsoonal climate that brings some moisture especially to the southern parts of the peninsula during the spring and early summer.

The entire Baja California peninsula is subject to rain-fall events related to the passage of hurricanes, often downgraded to the status of major storms after reaching landfall. Catastrophic flooding commonly results from such storms, which in Mexico go by the name *chubasco*. These are related to tropical storms that form southwest off the coast of Acapulco, typically during the months from August to October. For the most part, storm tracks start on a northward path along the Pacific shores of Mexico from below 15° N latitude, but usually veer off sharply to the west before reaching the southern tip of the Baja California peninsula at 23° N latitude (Romero-Vadillo et al., 2007). Between 12 and 20 such disturbances occur each season and the storms grow in intensity by feeding on warm ocean water as they move to the northwest. Those storms making landfall on the Baja California peninsula every few years have the potential to do serious flood damage. Calculations by Martínez-Gutiérrez and Myer (2004) found that rainfall dumped by

Hurricane Henriette over a 24-hour period in 1995 above Santiago in the southern Cape Region was capable of generating a discharge of ~132 m³/s from a drainage basin having an area of only 21 km². It was estimated that such a rate of discharge during a comparable time interval could result in the downstream transfer of sand and gravel in the amount of ~ 4,833 m³. Arroyos in this part of the peninsula eventually reach the Gulf of California at La Ribera, where the stream bed is unusually broad and filled with sand and gravel.

During recent years, as many as five tropical hurricanes made landfall on the tip of the Baja California peninsula (Fig. 9.2), causing major wind and flood damage. Three storms passed directly over Isla Cerralvo in the southern Gulf of California, where no one lives but the result of flooding from drainage basins smaller than the one above Santiago on the peninsular mainland carried similar bed loads of sediment seaward. Although down-graded from hurricane status after landfall, Marty and Ignacio in 2003, as well as John in 2006 and Odile in 2014 followed storm tracks that brought a significant amount of precipitation as far as Isla Angel de la Guarda in the northern Gulf of California. Normally dry for years at a time, arroyos on the eastern peninsular slopes that received precipitation from these storms all funneled water and sediment loads downstream toward the Gulf of California.

ENSO Cycles Present and Past

Warmer surface waters near the equator of the Pacific Ocean tend to pile up at times when atmospheric circulation becomes weakened and the convergence of trade winds pushing surface water westward fades in strength. One consequence of this phenomenon is that coastal upwelling off the coast of Mexico and Peru suffers a decline with an adverse effect on commercial fishing. This phase is part of a recurrent cycle called ENSO for El Niño Southern Oscillation that brings warmer weather and more rain to the Baja California peninsula and the Gulf of California. Associated with lower temperatures and a dryer climate, the La Niña phase of the cycle typically lasts from five to seven years.

Plate 1

Top: Isla San Luis (left), Cabo San Lucas arch (right).
Middle: San Basilio rhyolite (left), El Mangle limestone (right).
Bottom: Isla Coronados basalt (left), Isla Monserrat andesite agglomerate (right).

Plate 2

Top: Mulegé dune with wind etching (left), San Nicolás dunes (right).
Middle: El Rinchón shell beach (left), Monserrat shell beach (right).
Bottom: Punta Chivato and Islas Santa Inés (left), Isla Coronados (right).

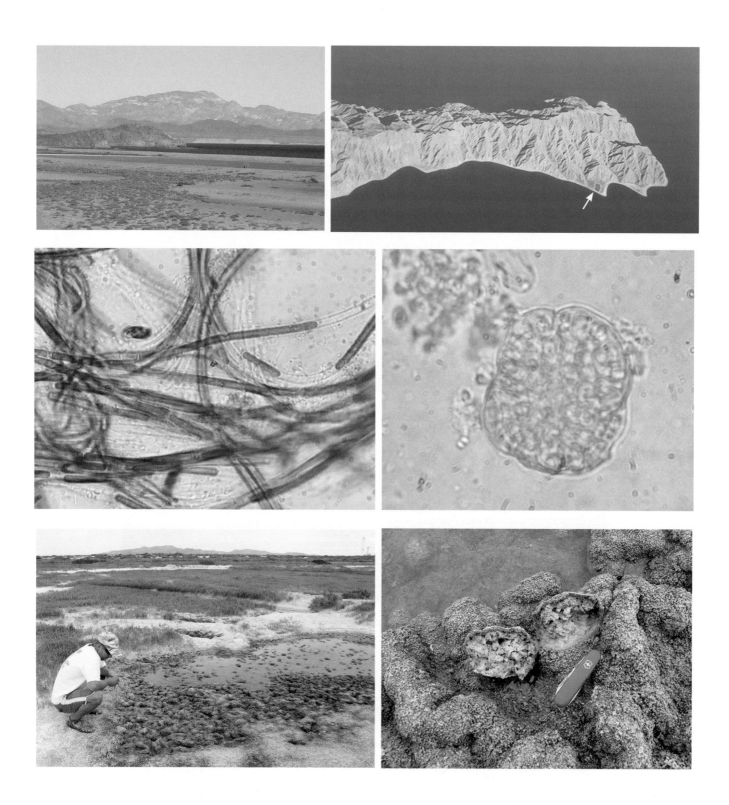

Plate 3

Top: Big lagoon Isla Angle de la Guarda (left), small lagoon (arrow) Isla San Lorenzo (right).
Middle: Bacterial filaments, 10 microns across (left), solitary bacterium, 30 microns (right).
Bottom: Punta Arena de la Ventana salt pond (left), bacterial dome (right).

Plate 4

Top: Isla Cerralvo fan deltas, numbered 1-6 (west shore).
Upper middle: Killer whales off south end of Isla del Carmen.
Lower middle: Sea lions on Las Galeras.
Bottom: Pliocene whale-bone fragments from Mesa Ensenada de Muerte.

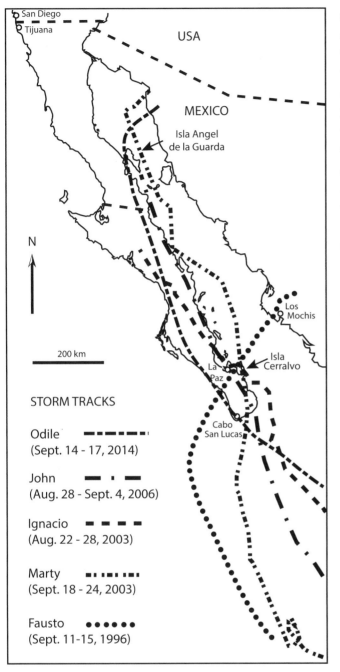

Fig. 9.2. Storm tracks from recent hurricanes striking Baja California and the Gulf of California.

The most severe hurricanes reaching Baja California have a tendency to occur during El Niño years, when the intensity of storms is heightend during the warmest phases of the ENSO cycle and hurricanes are more likely to enter the Gulf of California.

By use of drilling programs at sea, oceanographers have access to sediments that accumulate as layers on the ocean floor through long intervals of geologic time. Certain kinds of sediment rich in silica are derived from the plankton that thrive under conditions most favorable to upwelling across the equatorial zone of the world's oceans. At other times when upwelling is suppressed, different kinds of organic sediment are more likely to accumulate in the same zone. The start of ENSO cycles with wet and dry phases as understood today is linked to the onset of Northern Hemisphere glaciations during the Pleistocene more than 800,000 years ago. The glaciers that advanced and retreated in the far north during Pleistocene time had the effect of maintaining a stronger gradient in sea-surface temperatures from the equator to the North Pole. Based on information from deep-sea cores that span the boundary the Pliocene to the Pleistocene periods (Ravelo et al., 2004; Wara et al., 2005), detailed reconstructions of sea-surface temperatures and the depth of the thermocline along the equatorial Pacific Ocean suggest that conditions characteristic of a semi-permanent El Niño climate were dominant during much of the Pliocene. This climatic phenomenon is called the **Pliocene Warm Period**.

It is difficult to put a precise range of dates on the Pliocene Warm Period, but according to research on the history of surface-sea temperatures by Brieley et al. (2009), El Niño-like conditions were fully in place across both the mid-Pacific and mid-Atlantic oceans four million years ago. Today, the difference between sea-surface temperatures at the equator and at a latitude of 32° N typically is about 8° C (or a difference of about 14° F). During the early Pliocene, that difference is believed to have been only about 2° C (or roughly 2.5° F). Many who study climate change through geologic time argue that the early Pliocene from about five to three million years ago is the closest analog to the acceleration of global warming today. Under this interpretation, it is likely that hurricanes hit the Baja California peninsula and entered the Gulf of California more frequently in the past. At about the same time during the Pliocene, there is evidence for the retreat of fault scarps along the coastal plain (Mortimer and Carrapa, 2007) during a major interval of crustal uplift west of the Loreto region (Mark et al., 2014). Both activities had a tectonic source and both were sure to have an impact on bed rock exposure subject to stream erosion. During the subsequent Pleistocene, changes in global sea level associated with the waxing and waning of Northern Hemisphere glaciers also had an effect on the erosion of stream courses across the Baja California peninsula (Kluesner et al., 2014).

Modern Gulf Deltas and Estuaries

Numbered circles on the map in Figure 9.3 show the location of some outstanding examples of present-day delta systems and estuaries. Now largely inactive except for those occasions when substan-

tial amounts of river water are released from upstream dams, the Colorado delta (site #1) remains the largest feature of its kind on the Gulf of California. According to the classification scheme of Postma (1995), this huge system qualifies as a type-D delta.

Estuaries without prominent delta systems are denoted by the entrenchment of coastal inlets, where a mixture of fresh water and seawater is impounded. Known as *estuarios* in Spanish, three examples with brackish-water pools are found at El Rincón north of Mulegé (site #2, Fig. 9.4), at Mulegé (site #3, see Fig. 7.4), and at San Bruno north

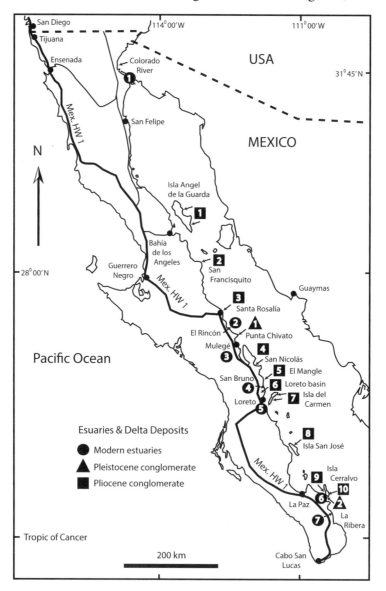

Fig. 9.3. Study sites for present-day estuaries and ancient delta systems in the Gulf of California.

Fig. 9.4. El Rincón estuary between Santa Rosalía and Mulegé.

Fig. 9.5. San Bruno estuary north of Loreto.

of Loreto (site #4, Fig. 9.5). The Mulegé estuary is partially blocked by the eroded hat-shaped structure of El Sombrerito (see Chapter 2). In the case of El Rincón, the deep *estuario* is 1.6 km long with embankments that rapidly rise to six or seven meters above sea level. It is connected through a complicated network of shallow arroyos across a dozen kilometers on the coastal plain to the confluence of two principal valley systems that exit highlands at an elevation about 200 m above sea level. In turn, those valley systems descend from source areas as much as 800 m above sea level. The San Bruno

estuario near Loreto also is well entrenched, but connected to a less complicated arroyo that crosses the coastal plain to the southwest for about 20 km where it reaches the escarpment of the Loreto fault at an elevation nearly 300 m above sea level. Farther inland above the fault trace, two valley systems converge with stream beds that descend from source areas close to 700 m above sea level.

The central trunk of the Loreto Arroyo enters the town of Loreto crossing a coastal plain 10-km wide (site #5, Fig. 9.3). La Higuera, Las Vír-

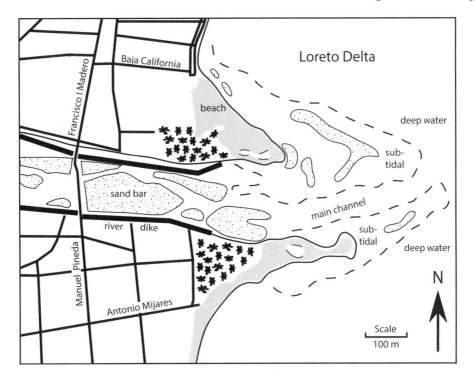

Fig. 9.6. River channel, beaches, and sand bars on the Loreto delta.

genes, and Las Parras are the three arroyo systems that converge on the Loreto Arroyo below the trace of the Loreto fault at an elevation about 100 m above sea level. The channel length measured from the delta inland to the source of the northern branch (La Higuera) is the longest at 22 km. At a distance 175 m from the delta front, the width of the arroyo channel is 175 m. The delta has two main lobes adjacent to the main channel, where it exhibits a classic fan shape spread over an area of 250,000 m² (Fig. 9.6). The levees near the opening to the Loreto delta are artificial and normally the channel of the Loreto arroyo would be unconstrained from flooding. During low tide, multiple bars composed of sand and gravel are exposed along the outer margins of the delta system. According to the classification scheme of Postma (1995), this small system barely qualifies as a type-C delta.

Modern fan deltas mark the termination of as many as 39 drainage basins that empty around the periphery of Isla Cerralvo (Fig. 9.3, overall locality #6). The island has an area of 136 km², making it the fifth largest in the Gulf of California. Cerralvo's arroyos may be differentiated according to length and average inclination (Backus et al., 2012). Many on the east side are short and steep with a maximum slope of 27°. Those on the western side tend to be longer and less steeply inclined. Individually, the fans are small because they represent the erosional product of small watersheds mostly less than 5 km² in size. Few fan structures exceed 500 m in width across the delta front, but they are distinctive and readily spotted on an over-flight (Plate 4, color inset). Inclination of the fan deltas is hard to calculate, because they tend to be truncated by wave action (Fig. 9.7). However, fan slopes appear to vary between 5 and 13° and correspond to type-B deltas (Postma, 1995). The Cerralvo deltas tend to be distorted in shape, due to strong long-shore currents driven by seasonal winds out of the north that transport sand down coast. Delta composition includes a range of clast sizes dominated by coarse silica sand derived from weathered granite, but also the cobbles and small boulders eroded from granite, basalt, and gneiss. The number of active fan deltas arrayed around the island is directly related to the

Fig. 9.7. Small fan delta on the west coast of Isla Cerralvo.

frequency of hurricanes that bring heavy rainfall to the area on a regular basis. The historical record of storm paths for named hurricanes between 1996 and 2014 (Fig. 9.2) shows that the island is vulnerable to direct hits.

Farther south near La Ribera (Fig. 9.3, site #7), arroyos Las Pocitas, Surgidero, and La Trinidad converge on a major arroyo that subsequently diverts into three distributary channels on approach to the Gulf of California. Inland 20 km near Santiago, the transport of fluvial sediments in the middle branch has been quantified and correlated to intense rainfall from hurricanes (Martínez-Gutiérez and Myer, 2004). Surprisingly, there is no vestige of a fan-shaped delta at the mouth of the system although the widest expanse between outer distributary channels amounts to 3 km. The narrow marine shelf at La Ribera drops rapidly into water 350 m deep 3.5 km offshore at the start of the Pescadero Canyon first mapped by Shepard (1950, Chart 9), which achieves a depth of 1,300 m within a distance of 12 km from the estuary. Although the amount of fluvial sediment carried by the arroyo system is enormous, it appears that the nearly full sediment load reaching gulf waters is flushed down a substantial marine canyon.

Geological Record of Gulf Deltas

Evidence for former deltas that merged on the Gulf of California is surprisingly widespread over much of the gulf's axial length. Only two examples of small Pleistocene deltas are marked on the map by numbered triangles, but they are separated by a distance of about 200 km (Fig. 9.3). Ten examples of Pliocene delta deposits are marked on the map by numbered squares. Many of the Pliocene examples are characterized by thick deposits formed by cobbles and boulders eroded from igneous rocks. The most northern examples from Isla Angel de la Guarda are separated from the most southern localities on Isla Cerralvo by a linear distance of about 700 km.

Pleistocene relationships. At Punta Chivato (Fig. 9.3, site Δ#1), a coral reef dated to the last interglacial epoch about 118,000 years ago sits partly above the toe of a deltaic deposit formed by river gravel that accumulated on the south flank of the Punta Chivato promontory (see Chapter 3). The deposit is composed of pebbles and small cobbles eroded mainly from andesite. These igneous clasts are smoothly worn and imbricated, showing internal organization (Johnson et al., 2007). **Imbrication** is the result of a strong current that causes flat pebbles and cobbles to overlap against one another in a more stable position like roofing tiles. Such a relationship is typical for clasts that come to rest after being carried by a strong current. The reef and underlying gravel bed is dissected by a modern arroyo. Test pits excavated on opposite sides of the arroyo suggest that the reef and associated delta was at least 100 m wide.

A succession of coral reefs (also from the last interglacial epoch about 122,000 years ago) is exposed in cliffs at the southwest end of Isla Cerralvo (Fig. 9.3, site Δ#2), where each of five coral beds sits on a pavement of conglomerate (Tierney and Johnson, 2012). In this case, the conglomerate was recycled from a nearby delta that emerged on the coast at the mouth of a Pleistocene arroyo. Unlike cobbles from river deltas, nearly all the clasts from the conglomerate beds associated with the coral reefs are encrusted with coralline red algae. Remnants of the actual delta are left intact only 200 m north from the reef deposit. Observed during low tide, the maximum preserved width of the

Pleistocene fan delta is about 100 m. Composition of the delta edifice includes cobbles and small boulders eroded mainly from coarse-grained granite.

Miocene-Pliocene relationships. The eastern flank of Isla Angel de la Guarda stretches over a linear distance of 70 km from the northwest to the southeast and features extensive fluvial and deltaic deposits of Pliocene age eroded from the island's western highlands (Fig. 9.3, site □#1). Reaching elevations in places more than 800 m above sea level, the highland's dominant rock is andesite that belongs to the Miocene Comondú Group. Anderson (1950, p. 38-42) crossed through the middle part of the island and described volcanic gravel and sandy deposits that are cross-bedded. Some of these fluvial deposits end in thick conglomerate beds exposed several places along the east coast, later mapped as such by Gastil et al. (1971). In studying the closed lagoons at the southern end of the island with their extensive mats of living stromatolites (Johnson et al., 2012), it was realized that the massive berms enclosing those lagoons are derived in part by gravel and cobbles reworked by present-day shore erosion (see Chapter 8). Replenishment of water in the closed lagoons depends in large measure on runoff from rainstorms that reach the island with the aftermath of hurricanes (Fig. 9.2). It is clear that the same process was effective during Pliocene time, when massive amounts of sediment were stripped from the highlands during floods.

The San Francisquito basin to the south (Fig. 9.3, site □#2) represents a more accessible locality, where a Pliocene delta easily may be observed. A distinctive channel from a paleoriver and its delta are preserved intact in the northwest corner of the basin (Johnson and Ledesma-Vázquez, 2009). As the upsteam drainage area of the paleoriver is small, the related delta also is relatively small. A modern arroyo, which now drains the same upland area, cuts a small canyon through the Pliocene delta deposit. Conglomerate with a thickness of one meter caps the top of the delta deposit (Fig. 9.8). It is nothing like the rocky-shore conglomerate rimming much of the San Francisquito basin (see Chapter 2), because here the granite cob-

Fig. 9.8. Granite cobbles from a Pliocene river delta at San Francisquito.

Fig. 9.9. Thick conglomerate beds from Upper Miocene strata at Santa Rosalía.

bles are partly imbricated and in direct contact with one another unlike rocky-shore conglomerate with clasts that are random in orientation and separated from one another by a sandy matrix. Although small in size, the San Francisquito paleodelta is eloquent for the quality of its geographic context that evokes a distinctly high-energy phase of Pliocene flooding.

The copper mining district at Santa Rosalía (Fig. 9.3, site □#3) sits within a basin formed in response to initial rifting in the Gulf of California during the late Miocene. A succession of terrestrial and marine formations filled the basin, which covers an area both north and south of town and extends 10 km inland (Wilson and Rocha, 1955). Exposures north of town along a major ridge incised by the Arroyo del Purgatorio reveal the basal part of the Boleo Formation with massive conglomerate and minor sandstone units exceeding 200 m in thickness (Fig. 9.9). The conglomerate is due to extensive erosion of the highlands to the west due to runoff that brought coarse clastics to a delta system covering as much as 30 km². Volcanic tephra trapped within the Boleo Formation allowed a date 6.76 million years to be determined (Holt et al., 2000). Succeeding the conglomerate are sedimentary rocks belonging to the Gloria and Infierno formations dated by fossils as Pliocene in age.

At San Nicolás (Fig. 9.3, site □#4), located 120 km southeast from Santa Rosalía, deltaic conglomerate beds are described from part (Los Volcanes Member) of the Pliocene San Nicolás Formation (Ledesma-Vázquez et al., 2006). The member includes various facies, one of which is represented by conglomerate distinguished by the close packing of clasts that range upwards to 26 cm in diameter derived mainly from andesite. Individual beds within the conglomerate tend to be massive with an average thickness of 5 m. The conglomerates are interpreted as fan deltas that spilled eastward into small basins that subsided as half grabens. It is significant that shell beds and related bioclastic sand (El Saucito Member) are interpreted as having accumulated offshore as shoals on the crest of tilted fault blocks isolated from the influence of deltaic clastics derived from highlands farther to the west. Based on analysis of the San Antonio tuff low in the San Nicolás sequence, the basin was active around 3.3 million years ago.

Located 28 km north of Loreto at El Mangle (Fig. 9.3, site □#5), it is possible to distinguish thin conglomerate beds deposited against Pliocene rocky shores from beds with closely packed andesite pebbles and cobbles characteristic of a minor delta deposit (Johnson et al., 2003, Appendix 1, Hotel section). Based on correlation with volcanic ash captured in the sequence nearby, stratification at El Mangle occurred about 3.3 million years ago close to middle Pliocene time. The scenario is remarkably similar to a modern pocket beach at El Mangle that is dominated by a wedge of andesite cobbles washed out from an adjacent ravine.

The Loreto basin is situated immediately north of Loreto (Fig. 9.3, site □#6), where Pliocene conglomerate, sandstone, and limestone layers are well exposed in the canyons eroded by arroyos San Antonio, Arce and Gua. Deltas in which foreset beds are strongly developed at the expense of topset and bottomset beds (**Gilbert-style deltas**) have been intensely studied in this area (Dorsey et al., 1997; Dorsey and Kidwell, 1999; Mortimer and Carrapa, 2007). The thickness of individual foreset units is shown to be as much as 50 m (Dorsey and Kidwell, 1999). Earthquake activity and subsidence on fault scarps initiated about 3.5 million years ago during the middle Pliocene was a formative process in the development of this extensive delta system. Between events triggered by earthquakes, banks composed of conglomerate and shelly sandstone accumulated in place with little sign of disturbance or layering due to the absence of active currents.

Separated from the peninsular mainland by the Carmen Passage with water that ranges from 390 to 460 m in depth, Isla del Carmen is the fourth largest island in the Gulf of California with an area of 143 km² (Fig. 9.3, site □#7). Two massive deposits of Pliocene conglomerate stand out among the many unusual geological features found on the island. Formally, they are called the Tiombó conglomerate and the Perico conglomerate (Johnson et al., 2016). A saddle or low-lying plateau formed by sedimentary strata occurs in the south-central part of Isla del Carmen reaching an elevation of about 100 m. Crowned by andesite rocks, higher ground to the south and in the north-central part of the island rises between 200 m and more than 400 m in elevation. The larger of the two conglomerate formations is the Tiombó conglomerate, which accounts for part of the island's distinctive plateau. The Arroyo Blanco limestone lies adjacent to the Tiombó conglomerate on its south flank, with more than 60 m of limestone beds dominated by rhodolith debris (Eros et al., 2006). It is the largest and chronologically most continuous body of limestone in the Gulf of California (see Chapter 5), and also occupies as much as half of the island's low-lying saddle. The Perico conglomerate is exposed in coastal outcrops a short distance north of Punta Perico in the northeast part of the island.

The contents of the Tiombó and Perico conglomerates were first mentioned by Anderson (1950), who described them as thick accumulations of "volcanic gravels." Both deposits are dominated by andesite rocks eroded from the Miocene Comondú Group, but the term gravel is misleading. An example from the lower part of the Tiombó conglomerate on the east coast of Isla del Carmen (Fig. 9.10) demonstrates that clast size ranges upward from cobbles to large boulders. Here, the accumulation is viewed as massive without any sign

Fig. 9.10. Massive conglomerate beds from the Pliocene delta near Arroyo Blanco on the east side of Isla del Carmen.

Fig. 9.11. Coastal exposure on the east side of Isla del Carmen with a 60-m thick succession of conglomerate beds.

of internal bedding. Individual clasts are packed together in close contact with little other sediment filling the spaces between. From a wider perspective, the most striking aspect of the Tiombó conglomerate is the arched profile shown by massive beds that extend in length over a distance of 2 km (Fig. 9.11). Limestone layers cap much of the Tiombó conglomerate as an extension of the Arroyo Blanco limestone that abuts the south side of the structure. At the northern end of the structure, the basal beds of the Tiombó conglomerate can be seen to sit directly on the basement andesite of the Miocene Comondú Group. Perpendicular to the shore, shallow valleys more than 50 m above the shore cliffs lead inland at the top of the structure. They afford places where the capping limestone and some of the underlying conglomerate beds can be observed to dip seaward to the west at about 5°. Based thus on layers that incline in three directions, the three-dimensional configuration of the conglomeratic body corresponds that of an enormous fan delta (Fig. 9.1).

The age of the Tiombó conglomerate is constrained by fossil pectens preserved in thin deposits from the lowest exposed parts of the structure, as well as other fossils that range through the thick Arroyo Blanco limestone that abuts the structure's south flank. Generally, the mega-delta complex built by the Tiombó conglomerate is roughly middle Pliocene in position correlative to the Piacenzian Age. The complex represents a type-B, tide-water delta after the criteria of Postma (1995) with a broad channel that strikes from west to east across the island. At first appearance, the only local source for the andesite materials incorporated by the delta would seem to be the highlands immediately to the north. Materials could not be imported from the south, because the Arroyo Blanco limestone fills a void between the delta complex and higher ground in that direction.

Considering the size of the Pliocene mega-delta, it appears to be much too large for the island on which it now resides. By comparison, remains of the Tiombó delta front are at least three times as wide as the present-day Loreto delta front. The ramifications of this comparison are that a former land bridge once existed linking the island to the mainland, and that a much larger drainage basin was responsible for the developement of the ancient delta front. The modern Loreto delta is fed by the converging branches of three upsteam arroyos with the total drainage area of 125 km². A map in-

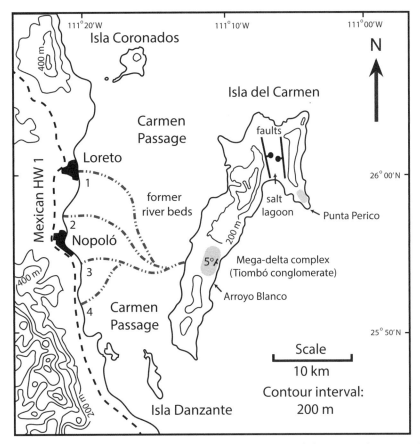

Fig. 9.12. Map reconstruction showing the Pliocene mega-delta complex on Isla del Carmen connected across a former land bridge with the present-day Loreto delta (1) and other coastal outlets for arroyos: El Tular (2), El Zacatan (3), and La Santa Maria (4).

terpretation for a mid-Pliocene land connection between Isla del Carmen and the peninsular mainland implies that the Tiombó mega-delta was fed by a much larger drainage basin incorporating other stream beds that now empty around the Nopoló area (Fig. 9.12). If correct, the present-day water shed that feeds the Loreto delta may represent only 30% of the mid-Pliocene water shed that ended at the Tiombó mega-delta prior to reactivation of faults responsible for submergence of the Carmen Passage (Johnson et al., 2016). On this basis, the name for the Tiombó mega-delta is derived from of a named segment of the Sierra de la Gigante ridge line high above Loreto to the west.

The geology of Perico ridge on Isla del Carmen (Fig. 9.12) is more complicated with regard to local faults (Dorsey et al., 2001) than the area surrounding the Arroyo Blanco limestone and Tiombó conglomerate. The Perico conglomerate may be as much as 750 m thick, although that thickness was estimated by taking into account strata offset by faults. Even so, a thickness of 450 m of the conglomerate is unambiguous and well exposed in sea cliffs north of Punta Perico. Sorting of the Perico conglomerate is considered poor to moderate with a content of andesite pebbles to boulders that attain a maximum diameter of 1.5 m (Dorsey et al., 2001). In contrast to the Tiombó conglomerate, better internal organization is implied by the concentration of larger clasts in certain intervals. The conglomerate on Perico ridge is entirely devoid of fossils. Also, the contact between the Perico conglomerate and older Miocene volcanic flows is better exposed, showing that the underlying surface of the volcanic rocks on which the conglomerate rests dips more steeply by a difference of 10° than the overlying conglomerate. Dorsey et al. (2001) concluded that the conglomerate formed as an underwater flow on a precipitous marine slope. In this case, the Perico conglomerate is consistent with the model for a "type A" deep-water delta (Postma, 1995).

Another observation on the Perico conglomerate is essential to the story as it relates to the question of tectonic influences on Isla del Carmen. Perico ridge is separated by a faulted valley from the most obvious source of eroded material in the island's northern highlands (Fig. 9.12). The south end of that valley is occupied by a salt lagoon that began to subside only during the Pleistocene (Kirkland et al., 1966). The reality of this relationship strengthens the proposal that earlier faulting on a larger scale was responsible for the decapitation of the Tiombó mega-delta from the peninsular mainland.

Located about 100 km southeast of Isla del Carmen, Isla San José is another place with potential for the study of Pliocene river and delta deposits (Fig. 9.3, site □#8). The complex basin tectonics on this island were studied by Umhoefer et al. (2007) and thick conglomerate and sandstone formations with affinities to river and delta deposits are known to be present. In particular, Pliocene strata derived from the erosion of granite are well exposed in arroyos on the east side of the island. The orientation of these strata is suggestive of foreset packages that prograded into the Gulf of California (Fig. 9.13). A closer view of thin layers within such a package shows that the sandstone is undisturbed and well sorted (Fig. 9.14), features typical of development in deeper water on a rapidly advancing delta front.

Described previously in this chapter with regard to modern and Pleistocene fan deltas, Isla Cerralvo also is a place where Pliocene fan deltas are well developed (Fig. 9.3, sites □#9 and □#10). In particular, the locality known as Paredones Blancos on the west coast of Isla Cerralvo is noted for a pair of massive conglomerate beds on the order of 10 m thick that sandwich an equally massive unit of limestone derived from crushed rhodoliths (see Chapter 5). The conglomerate beds at this locality (Fig. 9.3 site □#9) are exposed in vertical sea cliffs that extend along the shore with good lateral conformity for a distance of 0.75 km. A range of igneous and metamorphic cobbles and boulders are found to compose the conglomerate beds, including granodiorite, horneblend diorite, basalt,

Fig. 9.13. Overview of deltaic sand beds on Isla San José represented by Pliocene strata (background).

Fig. 9.14. Thin sandstone layers from the Pliocene delta complex on Isla San José (pocket knife for scale = 9 cm).

and foliated gneiss (Emhoff et al., 2012). Unlike the conglomeratic beds on Isla del Carmen, those at Paredones Blancos include a substantial amount of sand-size sediment derived from the erosion of granite. The source rocks for the Tiombó and Perico conglomerates on Isla del Carmen do not include granite or granodiorite. No marine fossils are present in the massive conglomerate beds at Paredones Blancos. The overall arrangement of strata at Paredones Blancos is interpreted as the result of changing sea levels that brought rhodolith debris into a flooded canyon mouth above a pre-existing delta deposit and reinstated a second delta front during a subsequent drop in sea level (Emhoff et al., 2012). A middle Pliocene age for the massive rhodolith bed is based on fossil sea urchins (*Clypeaster bowersi*) and pectens (*Argopecten revellei*) preserved as rare components of the limestone. In contrast to the scenario proposed for the Tiombó mega-delta on Isla del Carmen, the Pliocene deltas at Paredones Blancos were sourced exclusively from the island, because they partially fill a former canyon that leads to the island's interior.

A separate Pliocene delta occurs at Los Carillos on the southeast side of Isla Cerralvo (Fig. 9.3, site □#10). As mapped by Johnson et al. (2012), a conglomeratic detla deposit with a classic shape and dip shows an arcuate front extending for 200 m. Layers exposed within the delta sequence can be described as "volcanic gravel" but are very well cemented (Fig. 9.15). The layers also suggest grading by clast size similar to foreset beds in the modern Loreto delta. No fossils are present in the Los Carillos conglomeratic layers. As reviewed in Chapters 2 and 5, the adjacent rocky-shore conglomerate and associated limestone exposed to the south provides a marked contrast that includes fossil rhodoliths and other inter-tidal organisms attributed to the Pliocene (Johnson et al., 2012).

In summary, it is pertinent to restate that the Pliocene Epoch was an extraordinary window of time for the Baja California peninsula and Gulf of California. The peninsula's western rift margin concordant with the Sierra de la Gigante (and most probably other highland segments) experienced a major phase of uplift between 5.6 and 3.2 million years ago (Mark et al. 2014). The outcome of such

Fig. 9.15. Pliocene conglomerate from the delta deposit at Los Carillos on Isla Cerralvo.

uplift was an increase in the gradient over which highland slopes were susceptible to erosion from rainfall. Certainly by about four million years ago, the Pliocene Warm Period was in full swing (Brierley et al., 2009). This led to an intensification of hurricanes and major storms capable of striking the Baja California peninsula and all Gulf of California islands. Widespread occurrence of Pliocene delta deposits from Isla Angel de la Guarda in the north to Isla Cerralvo in the south suggests that the congruence of increased tectonic activity and a major shift in climate on the equatorial Pacific Ocean had a huge impact on the territory now belonging to Mexico's western frontier. Although there is evidence for Pleistocene deltas on the Baja California peninsula, they appear to be far less common and much smaller in size. Future field studies on the region's Pliocene and Pleistocene phenomena as reflected by geological formations of all kinds are needed to test this concept. In our own lifetime when concerns are raised about the rate of global warming, such studies are crucial.

Chapter 10

Coastal hydrothermal springs

Introduction

Prior to about 12 million years ago, the Gulf of California did not exist and what is now the Baja California peninsula was attached to mainland Mexico (see Chapter 1). Up until that time, dense oceanic crust (represented by basalt) off the west coast of Mexico pushed uniformly against and beneath the lighter crust of the North American continent (represented by granite). The process by which the denser crust is both pushed and pulled beneath continental crust is called **subduction**. Separation of the peninsula from the Mexican mainland began when the continental crust was stretched and fractured by extensional rifting. This process continued for some 8.5 million years until about 3.5 million years ago as the protogulf slowly opened in an east-west fashion (Ledesma-Vázquez et al., 2009). Rifting is associated with the initial subduction of the East Pacific Rise (main locus of rifting on the floor of the Pacific Ocean) beneath the North American continent. One of the regional consequences of such a rifting process is a thinning of the continental crust, where the Earth's geothermal gradient becomes accentuated.

Islands and some coastal hills inland along the western Gulf of California were elevated as rotated fault blocks (defined by **listric faults**) during this stretching process. These particular faults are often mistaken as normal faults at the surface, but at a greater depth they follow a curved zone of failure. Rocks at depth become hot, and the fractures associated with the faulting mechanism are unable transfer tectonic stress, because the confining rocks behave more as a plastic material rather than as solid rocks. Passages between the mainland and many of the associated islands formed when neighboring fault blocks (called **half-grabens**) slide and sank into the thinning crust to make depressions that later filled with seawater. The occurrence of

shallow, hydrothermal vents in the gulf region is directly related to the extensional tectonic regime, and more recently to transtensional movement along most of the pre-existing fractures. Hydrothermal activity is restricted to a network of regional listric faults that serve as conduits for the penetration of rainwater and seawater to depths where heat flow remains high.

Chemical Circulation Pathways

Along a descending pathway, faults allow for the circulation of meteoric water to reach a heat source at depth (Fig. 10.1). Because groundwater trapped in fault zones is warmer and therefore more buoyant, upwelling of hydrothermal fluids occurs along an ascending pathway from host rocks through the faults to the surface to issue as springs through permeable marine sediments, or directly into seawater at the seafloor. Quantities of gas including carbon dioxide and methane also may be infused into the water during this process.

Surface morphology provides clues to the mechanisms controlling fluid expulsion, particularly in the surficial expression of linear seeps aligned with faults (Forrest et al., 2005). Observed hydrothermal activity typically occurs sub-parallel to the onshore fault or fault system, suggesting that venting of the submarine fluid and gas is associated with local or regional fractures. For example, such a linear trend is found over a ~750 m distance near Punta Santa Barbara in Bahía Concepción (Forrest et al., 2005; Forrest and Ledesma-Vázquez, 2009), where side-scan sonar shows sinuous structures in sea floor sediment oriented subparallel to coastal faults. Copious discharge of geothermal liquid (90° C) and gas commonly occur in the intertidal and shallow subtidal zones along such structures. Escaping effluents may be traced to a depth of 13 m

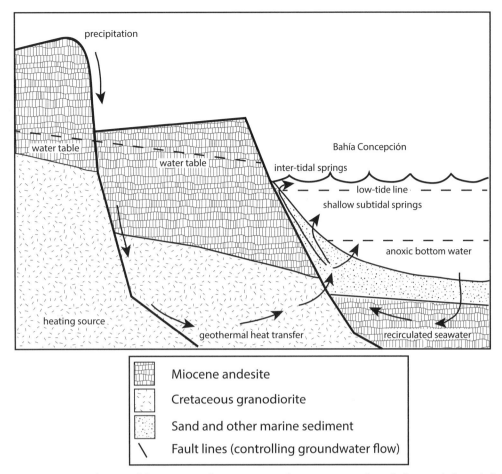

Fig. 10.1 Geothermal heat transfer to ground water circulated through local fault.

or more through soft sediments and fractures in rocks. Observations using SCUBA confirm that such structures are formed by elevated sediment compared to surrounding areas, where geothermal fluid and dissolved gas is vented. Hydrogeological studies conducted for the purpose of geothermal energy exploration (Arango-Galván et al., 2015), indicate that the primary permeability of rocks throughout the gulf region is very low, and entails the circulation of meteoric water in the process.

Hydrothermal activity shows a negative correlation with tidal height such that temperatures in the area of hydrothermal activity are as much as 8° C higher during low tide than at high tide. Gas samples reported by Forrest el al. (2005) show that the main components of the gas are N_2, CO_2, and CH_4, and that the methane is thermogenic in origin. Analyses of N_2 suggest it may be partially derived from the thermal alteration of algal material in immature sedimentary organic matter. One of the most interesting results from this research

(Forrest et al., 2005) is that He-isotope ratios show a significant mantle component (16.3%) in the gas. This signifies a direct link to deep-seated hydrothermal systems as found in the Guaymas Basin in the central Gulf of California, where the oceanic crust is relatively thin, and the mantle discharges hydrothermally enriched fluids directly into seawater. Mineralization of the chemical elements Mn, Ba, and Hg presently occurs in shallow-water settings as a result of hydrothermal venting (Canet et al., 2005). At some intertidal hot springs, moss-like crusts of manganese oxides and structureless detrital aggregates cemented by opal-A, barite, and calcite occur around the main conduits of discharge.

Influence on Marine Communities

The ecological food chain represents a general sequence of transfers of matter and energy in the form of food from organism to organism (see Chapter 11). The effluent from hydrothermal vents has a di-

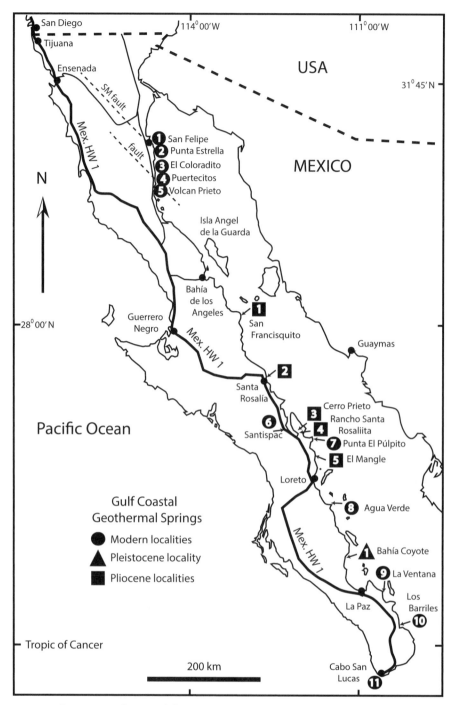

Fig. 10.2 Location of present-day and former geothermal springs in the Gulf of California.

rect effect on the biological communities living adjacent to them, as planktonic and benthic communities frequently found around the vents appear to be adapted to this type of habitat (Estradas-Romero et al., 2009). In general, high nutrient concentration stimulates high phytoplankton abundance. Although very small in size, **phytoplankton** are plants that convert solar energy to food by photosynthesis. They form the base of a marine food chain that leads farther to plant-eating animals and those that prey on the plant eaters. In the vent area, nitrate concentration typically is found to be higher than the surrounding area. Both in the vent area as well as the immediately surrounding area, concentration of ammonia tends to be higher than in normal seawater. This also is the case for sulphates and iron. The study by Estradas-Romero et al. (2009) indicates that diatoms are the dominant group affected, with a bloom of the species *Chaetoceros curvisetus*. Biological analysis of hydrothermal fluid also shows an abundance of cyanobacteria (including *Nostoc pruniforme* and *Trichodesmium erythraeum*). Phytoplankton data collected from different vent areas shows there is a significant difference in species composition from place to place, implying the presence of different communities. The observed differences in organism, concentration of chlorophyll-a, and the number of species observed can be related to the manifestations of hydrothermal discharge. Thus, it may be said that the region or areas associated with hydrothermal venting are a specialized part of the food chain fertilized by hydrothermal waters enriched in ammonia, sulphates, and iron.

Present-day Coastal Springs

The hydrothermal system widely in place along the western coast of the Gulf of California is shown by the accompanying map (Fig. 10.2) with localities marked by circled numbers (sites #1 to 10). All these sites are characterized by the recycling of local meteoric water and seawater heated from a deep-rock source (Fig. 10.1), as represented locally for the Bahía Concepción region (Santos et al., 2011). In addition to subtidal vents, there also occur subaerial springs that discharge onto the beach during low tide, which gives the name "Agua Caliente" (hot water) commonly applied to many such places. Obvious springs, such as the cluster of sites between San Felipe and Volcan Prieto (Fig. 10. 2, sites #1 to #5) are found in the Upper Gulf of California. Eocene to Oligocene sediments and conglomerates dominate the local geology, overlain by younger Miocene basalt flows. At San Felipe and Punta Estrella, the springs are accessible only during low tide but are directly linked with the junction of multiple, low-angle strike-slip faults generally related to the San Miguel Fault on the gulf coast (Barragán et al., 2001; Arango-Galván et al., 2015). At El Coloradito (site # 3), parallel faults cross through conglomeratic sandstone exposed in ridges at the seashore. The intertidal springs at Puertecitos (site #4) are well known to bathers. There, the thermal waters are saturated with sodium chloride and the water temperature rises above 70° C (Arango-Galván et al., 2015).

In the central part of the gulf region (Fig. 10.2, sites #6 to #8), the coastal geology is controlled by exhumed granitic basement rocks commonly overlain by andesite flows belonging to the Miocene Comondú Group (dominated by andesite). The hot springs at Santispac in Bahía Concepción (Fig. 10.3) occur behind of a dense mangrove forest (see Chapter 7), but are well marked and frequently visited by bathers. Other springs in Bahía Concepción are entirely subtidal (Forrest et al., 2005) with manifestations from 5 to 15 m in water depth. The temperature within bottom sediments at these sites may reach 87° C (Arango-Galván et al., 2015). Geothermal manifestations at Siquismunde are represented by hot springs and steaming ground that verge on the coastline at Punta El Púlpito (site #7), where a local landmark is formed by rhyolite thought to be only 500,000 years old. An intertidal hot spring off Agua Verde (site #8) related to fractures in the Miocene Comondú andesites is a favorite stopping place for boaters and kayakers going south to La Paz.

Fig. 10.3. Margin of the Santispac geothermal spring lined with rocks.

In the Lower Gulf of California that extends to the southern Cape Region, several areas with coastal hot springs are well known. Only a short distance north of La Ventana (Fig. 10.2, site #9), geothermal water seeps into the beach sands (Fig. 10.4) conducted by fractures in the underlying Cretaceous granodiorite. At Los Barriles (site #10),

101

Fig. 10.4 Geothermal spring source at the beach north of La Ventana.

the famous Buena Vista Resort gets its water from a geothermal spring that issues beneath the hotel and continues offshore (Forrest and Ledesma-Vázquez, 2009). Recent investigations of coastal hydrothermal systems have re-examined the network of WNW strike-slip faults in the Los Cabos block (Arango-Galván et al., 2015 and references therein). Chemical analyses of the coastal springs near Cabo San Lucas (site #11) suggest a strong recharge with seawater on account of the high sodium-chloride component to the thermal waters. The size of the shallow geothermal reservoir in this area is still under investigation, due to the preliminary nature of geophysical surveys.

Data summarized by Arango-Galván et al. (2015) on the active hydrothermalsystems throughout the Baja California Peninsula allow for the estimation of the reservoir parameters required to calculate the potential energy output for five subaerial and coastal geothermal prospects. In addition, the authors estimated the heat discharged by submarine systems as more than 6000 MWt as part of the region's computed geothermal resources. Those systems considered as most promising geothermal prospects are: San Felipe, Puertecitos, San Siquismunde, El Centavito and Agua Caliente, with a total combined estimated potential of ~ 400 MWe. According to Arango-Galván et al (2015),

this quantity would be sufficient to satisfy the region's increasing energy needs, especially taking into account that the geothermal resources are distributed all along the peninsula and may be utilized in areas not presently connected to the Mexican national electrical grid.

Examples of Former Geothermal Springs

Pleistocene relationships. Examples of Pleistocene geothermal deposits are uncommon around the Gulf of California. A single area at Bahía Coyote (Fig. 10.2, site Δ#1) has been adequately described in the published literature (DeDiego-Forbis et al., 2004). At this locality 70 km north of La Paz, paleohydrothermal activity is recorded along a narrow cliff within the coastal plain. Scattered spots in the immediate vicinity of small normal faults feature massive *Porites* coral beds and abundant molluscan fossils preserved in a sedimentary matrix stained green in color by material remobilized from the underlying Cerro Colorado. Typically, the corals are weathered, coated with a dark-colored patina, and poorly preserved. *Porites* fossils coated in silica are generally more common in the vicinity of the green-stained deposits. Based on field observations (De Diego et al., 2004), the affected areas are regarded as hydrothermal vents located along faults that served as active conduits for geothermal waters during the last interglacial epoch about 125,000 years ago. Fossilized remains of *Nassarius tiarula* are known to be abundant in the same surrounding area, which is interesting considering that present-day nassariid gastropods occur in abundance around shallow-water vents at Bahía Concepción (Forrest and Ledesma-Vázquez, 2009) and elsewhere throughout the world.

Pliocene relationships. Five examples are described in this section, the northern most of which occurs in the San Francisquito embayment (Fig. 10.2, site □#1). There, Pliocene strata occupy a 10-km² area located 75 km due south of Isla Angel de la Guarda. Large-scale relationships entrained in this landscape offer the opportunity to examine biofacies patterns and depositional processes on rocky shores both outside and within the Pliocene basin (See Chapter 2 on rocky shores). Only 400 m long, the north channel at San Francisquito is defined by a pair of faults forming a graben. Moreover, this particular feature follows lineaments parallel to transform fractures still active in the Gulf of California. Granodiorite hills rise on the west and east sides of the inlet to elevations between 100 m and 175 m. Remnants of a limestone ramp are preserved on the inner west side of the channel sitting on granodiorite. It is assumed that the same ramp formerly extended to the opposite shore. Among fossils preserved in the ramp is an echinoid (*Clypeaster revellei*) considered diagnostic for the middle Pliocene (Durham, 1950). Brownish-colored crusts of geothermally altered carbonates fill low areas eroded into the ramp at this locality.

A large Pliocene basin opens onto the wide beach at nearby Ensenada Blanco, anchored on the north and south by granodiorite hills. Within, a rocky shoreline traced by the Cretaceous-Pliocene nonconformity follows an arc 14 km in length (Johnson and Ledesma-Vázquez, 2009). The elevation at which the nonconformity is exposed varies due to fault relationships. Two boundary faults are associated with the basin, one of which extends along the north margin. The other runs along the inner east margin, but also cuts through limestone and granodiorite to reach the outer coast. These and other related normal faults buried below the basin facilitated erosion of a topographic depression prior to initial flooding in mid-Pliocene time. Faults on the half grabens show projections that intersect both entrances to the basin, which implies that transtensional tectonics prepared the way for flooding. Boundary faults inside the basin were active during flooding, but also remained active sometime afterwards. High-angle dips suggest that the ramp on the basin's north side was steepened after deposition, although contemporaneous steepening also occurred during ramp accretion based on changing dip angles preserved internally within. Additional evidence for reactivation of boundary faults comes from localized hydrothermal deposits draped along the upper margin of the ramp (Fig.

Fig. 10.5. Discolored crusts from geothermal springs, north side of the San Francisquito basin.

10.5), where brownish-colored crusts occur similar in composition to those in the adjacent north channel. More along the basin's west margin, fossil bivalves (*Glycymeris maculata*) are commonly nestled in growth position among cobbles and boulders. The same bivalve is associated with a series of three mounded hydrothermal vents having an alignment consistent with a buried fault below the northwest segment of the paleoshore (Fig. 10.6). Each with their own fossil association, the array of mounds linked by the same structural trend strongly sup-

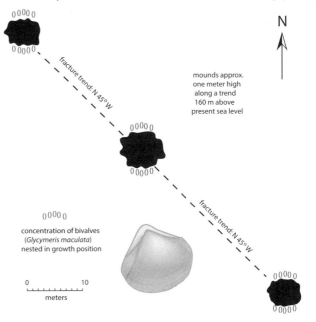

Fig. 10.6. Schematic diagram for a line of Pliocene geothermal springs with fossil content in the San Francisquito basin (after Johnson, 2014).

ports the interpretation that Pliocene marine shells thrived at these spots at a time when geothermal waters were leaked into the San Francisquito basin.

Farther south at Santa Rosalía (Fig. 10.2, site □#2) is an incipient rift basin that formed as a result of northwest-to-southeast trends in pre-gulf continental rifting during the Late Miocene. This basin hosts the famous Cu-Co-Zn mineralization from the Boleo mining district, as well as manganese-oxide deposits of nearby areas. Therein, a series of sedimentary marine and non-marine formations were deposited that include the Boleo Formation. Varying in thickness from 250 to 350 m, the Boleo Formation is divided into members that include a basal conglomerate followed by limestone and gypsum member, capped by clastic rocks. Mineralization consists of laterally extensive and stratiform ore bodies (called mantos) with disseminated Cu-Co-Zn sulfides and manganese oxides, constrained between fine-grained sediments at the base of each of various cycles within the Boleo Formation.

The NW-SE trending mantos constituted by manganese oxides and copper silicates cross-cut Miocene andesite rocks that represent the regional basement. These structures served as conduits for the ascent of the hot mineralizing fluids discharged into the Santa Rosalía basin. The fact that the mineralizing fluids ascended through these structures is inferred by the juxtaposition of high-grade Cu ± Co zones and localized discordant to stratabound zones of pervasive Mn-Fe-Si alteration (Del Rio-Salas et al., 2003). A range of temperatures between 18° C and 118° C is interpreted for the manganese oxides in the different Boleo mantos.

Geothermal waters from springs and wells in and around the Tres Vírgenes and La Reforma caldera fields located north of the Boleo basin are characterized by temperatures from 21 to 98° C. Inland close to El Púlpito (south of the Boleo district), localities such as San Saquicismunde and El Volcan are current examples of hydrothermal activity that may correspond in type to that responsible for the mineralization in the Boleo district (Arango-

Galván et al., 2015). These geothermal emanations are geographically quite dispersed, but share similar geological features to include a structural northwest-southeast control, evidence of hydrothermal alteration, occurrence within Miocene volcanic or volcanoclastic rocks, and a low-temperature range between 38 to 94 C°. Field observations and Pb and Sr isotope signatures show that the fluids involved in mineralization of Mn from the Boleo district are essentially hydrothermal and exhalative in origin by Del Rio-Salas et al (2003; and references within).

Bahia Concepción is an example of a region characterized by coastal hydrothermal venting in a fault-bounded bay. This large bay formed during the Late Miocene extension that affected the broader Gulf of California region (Ledesma-Vázquez and Johnson, 2001). Bahia Concepción is one of the largest fault bounded bays in Baja California, and its half-graben configuration developed during that interval. Several lines of evidence point to extensive paleohydrothermal activity along the Concepción Fault Zone, such as the manganese veins exploited at the Gavilán mine on Peninsula Concepción. Several faults cut across Pliocene strata parallel to the Concepción Fault Zone at the base of the Peninsula Concepción. These faults acted as conduits for hydrothermal fluids, resulting in paleohydrothermal vent sites. A paleohydrothermal vent site that occurs along a major fault line is located on the northeast side of Cerro Prieto (Fig. 10.2, site □#3), where red mudstone shows evidence of a peculiar disruption reminiscent of gas bubbles rising through fine sediments (Johnson et al., 1997). Preservation of this feature in Upper Pliocene strata means that fault-fed hydrothermal activity occurred during the late Pliocene. This site offers an excellent example of the "gasohydrothermal" activity now occurring at modern shallow-water hydrothermal areas in nearby Bahía Concepción.

Another remarkable site of fault-controlled paleohydrothermal activity occurs south of Cerro Prieto at Rancho Santa Rosaliita (Fig. 10.2, site □#4), featuring a member-level stratigraphic unit in the Infierno Formation designated as El Mono. This unit is represented by 15 m of brown or white chert interbedded with soft, poorly consolidated layers of finely brecciated chert (Ledesma-Vázquez et al., 1997). Individual chert beds within El Mono Member range in thickness from a few centimeters to a maximum of 4.5 m. Bioturbation is noticeable in some chert layers, where burrow systems are exposed (Fig. 10.7). Silicified mangrove rootlets found in the brecciated chert layers also confirm that deposition took place in shallow intertidal to upper intertidal waters. The in situ root casts belong to a species of the black mangrove (*Avicenia germinans*) and corroborate the deposit's coastal setting (see Chapter 7). A Late Pliocene age is indicated by the index fossil of a sand dollar (*Clypeaster marquerensis*) commonly found in the overlaying limestone unit of the Infierno Formation (Johnson et al., 1997).

The repetitive sequence of reworked and massive chert beds from El Mono Member formed in a shallow-marine environment most likely influenced by a high rate of evaporation. Evaporation would have increased water salinity, which in turn promoted silica precipitation. Salinity increased to the point that halite precipitated along with silica in some of the beds, but the basin did not completely evaporate, as indicated by the absence of saline crusts or salt deposits. However, the water was shallow enough to allow wave action and erosive

Fig. 10.7. Trace fossils in El Moro chert, Rancho Santa Rosalita on the Concepción peninsula.

processes that reworked some of the chert beds and introduced small amounts of terrigenous materials. About 30 km north of Loreto at El Mangle (Fig. 10.2, site □#5), a north-south oriented tectonic block named for that location (Johnson et al., 2003) resides on the east flank of the Cerro Mencenares volcanic complex. The block is configured as a 7-km by 1-km horst separated from Cerro Mencenares by a north-south valley fault that formed during the extensional phase of tectonics in the gulf sometime prior to about 3.5 Ma. The block's south end is defined by a normal fault striking N 55° W parallel to the Atl Fracture in the modern Gulf of California.

The diagonal El Coloradito Fault truncates an undisturbed carbonate ramp sequence continuous to the south, which includes a tuff bed yielding a K/Ar age of 3.3 ± 0.5 Ma (Johnson et al., 2003). The fault's Spanish name refers to the striking red coloration in coastal cliffs near Punta El Mangle that resulted from hydrothermal alternation. It appears that El Coloradito Fault was active sometime after the ash fall, but the overall timing fits with a change in regime to transtensional tectonics initiated in the Gulf of California roughly 3.5 million years ago. A succession of smaller faults that run parallel to El Coloradito Fault cut across El Mangle Block father to the north at close intervals and hydrothermal alteration is extensive along all these oblique faults. Thick beds with abundant fossil pectens drape parts of the block from elevations as high as 100 m down to sea level. It is noteworthy that fossil pecten beds are heavily tectonized and mineralized by coatings with a green discoloration, which owe their origin to hydrothermal activity.

The extensive carbonate ramp that borders El Mangle Block to the south includes a 1.8-m thick unit of layered opalite enriched by extensive debris of woody plant fragments (Fig. 10.8, see inset photo). The unit sits on weathered red clay, which in turn overlies andesite bedrock related to the Cerro Mencenares volcanic complex. The dense clay appears to have formed a water-tight boundary above which geothermal water enriched in silica circulated through sediment with abundant plant mate-

Fig. 10.8. Pliocene opalite beds with fossil plant debris (see inset) at El Mangle.

rial, some of which is almost certainly related to the same white mangrove species living at El Mangle today (see Chapter 7).

In summary, not only are modern geothermal springs extensively developed along much of the Baja California gulf coast, but many localities dating especially to the Pliocene also feature geothermal deposits with a surprising range of fossil inclusions. These contain both marine fossils (San Francisquito and El Mangle Block) and mangrove fossils (Rancho Santa Rosaliita and El Mangle) from the edge of open lagoon settings. At a more remote time dating back to the middle Miocene rifting of andesitic crust now located on the gulf margin of the peninsula, geothermal activity was responsible for emplacement of metals and sulfides in the mining district of Santa Rosalía. The potential for similar geothermal deposits in places south of Bahía Concepción toward Loreto is of keen interest to the mining community and recent activity has revolved around various concessions in that region. On a different level, the Mexican government through its Comisión Federal de Electricidad maintains an active interest in the utilization of geothermal power generation throughout the region.

Chapter 11

Denizens of the Open Sea

Introduction

The ecosystems treated in Chapters 2-10 of this handbook are situated directly along the coastline or within a few kilometers of the shore on the sea bed in water depths less than 50 m. This chapter examines connections between organisms living in the coastal ecosystems (typically marine invertebrates) and the larger more noticeable creatures (mainly marine vertebrates) that populate the open sea. The Gulf of California is a truly vast body of water with physical dimensions that are hard to grasp. Stretching from the delta of the Colorado River in the north to the opening with the Pacific Ocean in the south, the gulf measures 1,100 km in axial length. It has a maximum width of 180 km between the tip of the Baja California peninsula and mainland Mexico. Overall, the gulf's surface area amounts to 177,000 km² and this includes 2,850 km² occupied by more than 30 islands. The collective habitat space represented by coastal ecosystems is small compared to the territory of the deeper, open seaway.

Variations in water depth throughout the Gulf of California are extreme. In the northern part, water depth rarely exceeds 200 m as a consequence of the high sediment load formerly emptied by the powerful Colorado River prior to its obstruction by dams. Still in the far north, surprisingly deep basins occur between the peninsular shores and the San Lorenzo islands. The passage at that location is called the Salispuedes Channel. Its submarine topography was first explored in 1940 (Shepard, 1950) and found to register water depths between 1,250 and 1,450 m. The southern entrance to the Gulf of California features basins that are as much as 3,000 m deep. The denizens of the open gulf are blue-water creatures accustomed to a habitat that is essentially oceanic in nature as regards water depth.

Under the assumption that the shelf margin around the Gulf of California extends outward from the shore for an average distance no more than 3 km to reach the 50-m isobath (but also including those shelves around the gulf's many islands), the combined habitat area of coastal ecosystems amounts to little more than 5% of the gulf's water surface. Even so, links between the open sea and the more confined space of the coastal ecosystems are profound and bear examination in regard to the overall biodiversity of the complete gulf region. The pathways through which food circulates among all life forms in the Gulf of California makes an absorbing story. It is a story that has played and replayed in continuous action for no less than five million years.

Modes of Life and Sustenance

Whether animals or plants, all marine life is divided into groups based on basic modes of life. Fixed organisms that live most of their lives firmly attached to rocks, like barnacles or encrusting red algae, are regarded as sitters. Crawlers are those organisms that stay on the sea bottom, but move around from place to place. A crab is a good example of a crawler. Rhodoliths don't move by their own free volition, but do represent an unusual form of coralline red algae (see Chapter 5) that rolls on the sea floor under the influence of waves and bottom currents. A more formal designation for sitters and crawlers is that they belong to the **benthos** as bottom dwellers. Swimmers include creatures like shrimp or fish, spending most of the time moving through the water column. The technical designation for swimmers is that they belong to the **nekton**. Organisms that float and drift on the surface or maintain neutral buoyancy in the water column are regarded as part of the **plankton**. For the most part, single-celled plants populate the **phytoplankton**, whereas single-celled or very small animals belong to the **zooplankton**. The latter entails animals that have reached their full maturity, such as the tiny crustaceans called copepods, whereas other components of the zooplankton include the larvae of many other kinds of marine invertebrates and fish.

The ways in which life sustains itself give rise to another classification scheme. Life is divided into two camps: the producers (mainly cyanobacteria and plants) and the consumers (mostly animals). Plants and bacteria that generate their own food using sunlight through the process of photosynthesis are regarded as **autotrophs** (from Greek roots, meaning "self-feeding"). Primary consumers are the animals that obtain food directly from the producers, and they are better known as **herbivores** (plant eaters). Secondary consumers are those animals that get their food energy from the herbivores, and they are better known as **carnivores** (meat eaters). Some animals maintain themselves with a mixed diet of both plants and animals. They are regarded as **omnivores**. Plants and animals that manage to reach full maturity and expire without being eaten by primary or secondary consumers go through a process of decomposition. Another class or animals called the **detrivores** (eaters of organic detritus) depend on this material as a food source. In the marine realm, crabs and some sea snails are good examples of detrivores.

Food Pyramids, Chains, and Webs

In *The Log from the Sea of Cortez*, John Steinbeck comments on the natural fertility of a nudibranch (a kind of sea snail without a shell), also known as a sea hare. He observes that during a single breeding season, a single female may produce as many as 478 million eggs, and he argues that the world would rapidly become overwhelmed by that species if all those eggs grew into mature animals that continued to reproduce. Clearly this is not the case. The eggs of the nudibranch *Tethys* enter the marine realm as part of the zooplankton, and they provide a rich source of food for other animals to feed on. Relatively few fertilized eggs survive to insure the survival of any particular nudibranch species. What is true for the sea hare applies also to most other forms of life in the Gulf of California and all around the world both on land and at sea.

The concept of over production in the reproductive cycles of many organisms leads naturally to the notion of a tiered food pyramid (Fig. 11.1). Essentially, the feeding potential of primary producers is enormous as supplied by phytoplankton and benthic algae. In scope, the volume of food available from these sources to first-order consumers in the next tier of organisms provides the base of a massive food pyramid. For example, many clams (bivalves) are filter feeders that pump large volumes of seawater through their gills and subsequently harvest food from the abundant phytoplankton. The chocolate shell (*Megapitaria squalida*, see Chapter 4) is a good example of an infaunal bivalve that hides out of sight below the sand surface and uses siphons to draw food-laden seawater into its shell. Many sea snails (gastropods) also are herbivores that feed on marine algae. For example, the turban shell (*Turbo fluctuosus*) is a common intertidal grazer that feeds on microscopic algae using a mineralized mouthpart (called a **radula**) like a file to rasp algae from the rocks where the algae grow in profusion.

Based on current surveys of the principal mollusks living in the Gulf of California (Hendrickx et al., 2005 and Brusca and Hendrickx, 2010), there are 566 species of bivalves (most of which are filter feeders) and 1,534 species of gastropods (only some of which are grazers). Collectively, the diverse populations of clams and sea snails provide a substantial food source for the next tier in the food pyramid, represented by first-order carnivores. A good example of a predator from the invertebrate world that typically feeds on clams and sea snails is the starfish, and a good example of an important vertebrate animal that utilizes the same food source is the stingray (Fig, 11.1). Some of the specialized anatomical characteristics that make these particular animals so efficient as hunters are described in a following section (below). The analogy with a pyramid is a useful way of thinking about scaled

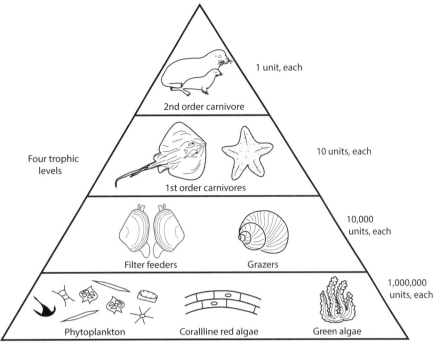

	1 unit, each
2nd order carnivore	
Four trophic levels	1st order carnivores	10 units, each
Filter feeders Grazers	10,000 units, each
Phytoplankton Corallline red algae Green algae	1,000,000 units, each

Fig. 11.1. Model food pyramid for the Gulf of California.

predator to prey ratios, and it is exemplified by placing a mammal like the California sea lion (*Zalophus californianus*) in a higher position as a second order carnivore. In fact, a food pyramid may include a succession of third, fourth, and even fifth order carnivores, but the object of the exercise is to illustrate a progression of trophic levels in which the population of organisms in any given tier far surpasses the number of organisms reliant on that tier as a food source.

Within any model for a food pyramid, a variety of food chains are readily apparent. A single pathway illustrated in Figure 11.1 flows from phytoplankton at the base to infaunal clams and onward to smaller stingrays preyed upon by sea lions at the top of the chain. Inherent in any open marine setting, there are multiple food chains that are complex and inter-woven with one another to make a more intricate food web (Fig. 11.2). Such a food web for the Gulf of California starts with the same broad base built on phytoplankton, but leads onward to zooplankton that are fed upon by many different kinds of fish, which are directly preyed on by pelicans from the air or by even larger fish like the blue marlin that are masters at attacking schools of fish. The Gulf of California waters are famous for the great fish boils that surge with tre-

mendous energy to attract pelicans, sea lions, and other predators feeding at the margins. At the top of such a food web is an apex predator like the killer whale, or orca (*Orcinus orca*), that lacks any natural predator of its own. Orcas are transient visitors to the Gulf of California (Plate 4, color insert) but do appear from time to time in small pods typically led by females. The food chain that ends with the orca is long, varied, and complex. It may terminate with spectacular acts in which the orca engages prey items like a young sea lion. At the other extreme, a food chain may be short, but stupendous in scale, as represented by a marine mammal like the blue whale (*Balaenoptera musculus*), the world's largest animal and also a transient visitor to the Gulf of California, feeding directly on zooplankton without any other links in the chain. Other whales that feed directly on the zooplankton, such as the humpback whale (*Megaptera novaeangliae*), are more common and spend more time in the Gulf of California.

Invertebrate Predators

The habitat space limited to coastal ecosystems is small compared to the open Gulf of California, but predation on marine invertebrates by other marine invertebrates is intense within these systems.

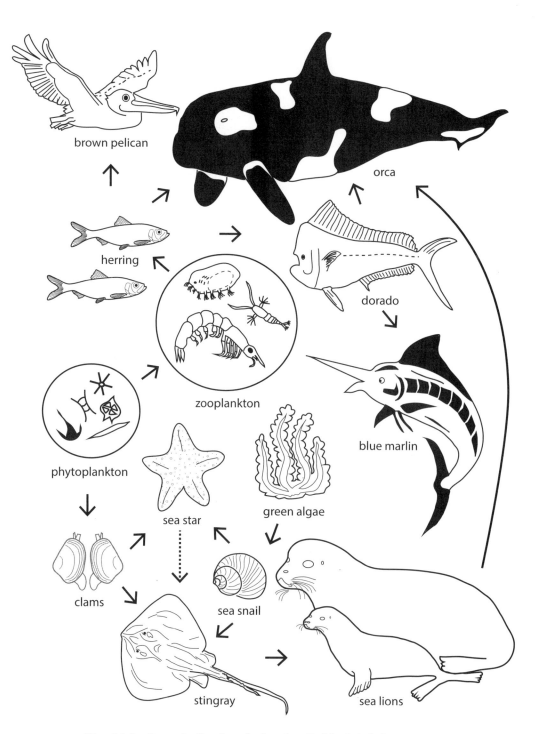

Fig. 11.2. Sample food web for the Gulf of California.

Stony and Soft Corals. The stony coral (*Porites panamensis*) with finger-like branches that forms the principal component in reefs around Isla Espíritu Santo and Cabo Pulmo (see Chapter 3) ranks as an important first-order carnivore preying on zooplankton. The soft coral (*Palythoa* sp.) that lies flat like a mat exhibits much larger polyps in its colonial assembly. These larger polyps are capable of feeding on larger prey items, such as small crusta-ceans and small sea snails. Both the stony and soft corals use stinging cells (cnidocytes) attached to flexible tentacles to stun prey and convey the prey to the mouth.

Gastropods. Whereas some sea snails are passive grazers, others are highly efficient predators. The cone shell (*Conus princeps*) lives in the intertidal zone under rocks and is easy to identify due to its blunt conical shape (Fig. 11.3) and orange-pink coloration with many dark axial lines. Equipped with venom glands and a harpoon-like radula that punctures prey, the cone shell takes other mollusks, marine worms, and even small fish. The thais shell (*Neorapana muricata*) also lives in the inter-tidal zone, generally on rocks. Strong nodes that protrude above the body whorl and spire (Fig. 11.4) help to identify this species, which employs a mineralized radula in its mouth parts to bore through the shells of bivalves and small crustaceans. Digestive juices help to dissolve the prey's tissues in place. With its distinctive cylindrical shape, the olive shell (*Olivia davisae*) is an infaunal sand dweller that lives in the shallow sub-tidal zone. Sensing prey represented by other mollusks and small crustaceans through chemical smell, it comes to the surface and engulfs its prey with a comparatively large foot. In the process, the prey is smothered in slime excreted by the olive shell, and is carried below the sand surface for consumption.

Cephalopods. Twenty species of cephalopods (including the octopus and various squids) are known from the Gulf of California (Hendrickx et al., 2005 and Brusca and Hendrickx, 2010). For example, the two-spotted octopus (*Octopus bimaculatus*) is a small benthic animal that measures up to 18 cm in length. It dwells in the intertidal zone to a depth of 50 m. Eight tentacles surround the mouth, each of which is armed with suckers. Suckers take a firm hold on a prey item and the tentacles convey the prey to the mouth, which is equipped with a strong, horny beak. The beak is applied to bite and inject venom into prey. Favorite food items include crustaceans and other mollusks.

The paper nautilus or argonaut (*Argonauta corniculatus*) belongs to a group of octopi specialized for full-time activity in the water column near the surface. Differences between male and female argonauts (**sexual dimorphism**) are very strong. The male is much smaller than the female and the larger female produces a "paper-thin" calcareous egg case that is loosely coiled in a flat-spiral form (Fig. 11.5). Pair of specialized dorsal tentacles is used to secrete the shell. After depositing her eggs

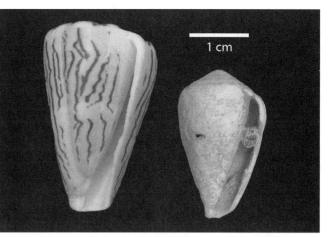

Fig. 11.3. Modern and Pleistocene cone shells (*Conus princeps*) from Punta Chivato.

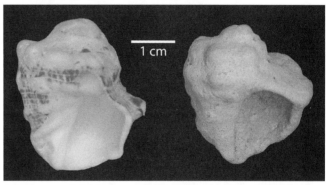

Fig. 11.4. Modern and Pleistocene thaid shells (*Neorapana muricata*) from Punta Chivato.

Fig. 11.5. Shell of the paper nautilus (*Argonauta corniculatus*) from Punta Chivato.

within the case, she takes up occupancy typically with the head and tentacles protruding from the shell. Unlike the pearly nautilus from the western Pacific Ocean, the paper nautilus lacks a chambered shell. Although some buoyancy is maintained due to air trapped within the shell, it is used mainly as a brood chamber for eggs. Like other octopi, the paper nautilus has eight arms with suckers and a mouth equipped with a biting beak. Feeding occurs during the daytime and favored prey items include small crustaceans, mollusks, and jellyfish. In turn, the paper nautilus is preyed on by the blue marlin and dolphins. Like other cephalopods, the paper nautilus avoids attack by emitting a cloud of dark ink.

Sea Stars. Echinoderms (invertebrates with a spiny skin) from the Class Asteroidea include the carnivorous sea stars. Tube feet on the underside of the arms are used to apply pressure by suction to open clamshells. An eversible stomach is extruded from the body and inserted through even the smallest gap to penetrate inside the valves of clams. Stomach acids dissolve the prey in place. In the Gulf of California, 63 species belong to this class (Hendrickx et al., 2005 and Brusca and Hendrickx, 2010). As an example, the sun star (*Heliaster kubiniji*) is a multi-armed asteroid among the most common in the region. It begins life with five arms, but adds more through time and typically has from 19 to 25 in adult life. This species dwells in the middle to lower intertidal zones and feeds on epifaunal mussels (like *Modiolus capax*) and barnacles. Another example is the Panamic cushion star (*Pentaceraster cumingi*), easily recognized by the blunt, red spines that cover its upper surface. This sea star exhibits five arms throughout its life cycle. It lives in the lowest tidal zone to a depth of about 10 m and also feeds mostly on mollusks. Few animals prey on starfish due to their tough skin, but the manta ray is known to feed on them.

Vertebrate Predators

Fish. The diversity of fish species within the Gulf of California was recognized to be very high from the earliest organized surveys beginning in the late nineteenth century. Large predatory fish, like the dorado (*Coryphaena hippurus*) and blue marlin (*Makaira nigricans*) were already known during the early 1800s. Today, the region's confirmed number of boney fish species is 821 (Hastings et al., 2010). Particularly among the smaller fish, additional species are sure to be discovered and added to this list in future years. The tally is complicated by geography, because the fish fauna includes a mixture of both temperate and tropical species. Overall, the fish fauna is dominated 87% by tropical fish that also occur outside the gulf to the south. A smaller percentage of temperate fish is known to live outside the gulf to the north. About 10% are endemic fish that live exclusively within the Gulf of California (Hastings et al., 2010. Overall, the diversity of the fish stock increases from north to south, with approximately twice as many species living in the far south as in the far north.

Fish stocks like the herring (*Harengula thrissina*), occur in large schools over soft bottoms and feed on the abundant zooplankton (Fig. 11.2). These, in turn, are preyed on by a succession of larger fish concluding with a species such as the blue marlin. As discussed previously (Chapter 5), the coastal rhodolith banks and mangrove forests (Chapter 7) serve as key umbrella ecosystems in which many fish species are supported especially during their early growth forms. Biodiversity surveys have found that as many as 33 fish species are commonly associated with rhodolith banks, as for example in Bahía Concepción (Foster et al., 2007).

Sharks and rays. These predators are classified among the "fish" but treated here under a separate category owing to their specialized anatomy and feeding patterns. Examples from the Order Chondrichthyes include the white shark (*Carcharodon carcharias)* and the Pacific manta ray (*Manta hamiltoni*), both of which make appearances in the Gulf of California. The manta ray feeds on zooplankton by taking in large amounts of water while swimming with an open mouth. There are 87 different species of sharks and rays registered as living in the gulf region, either in permanent populations or as transient visitors (Hastings et al., 2010).

The round stingray (*Urobatis halleri*) is a common resident in coastal gulf waters, where its broad, flat body shape and the extension of a short tail armed with a long, serrated stinging spine is readily recognized. This species feeds on marine worms and the abundant supply of mollusk clams and sea snails (Figs, 11.1 and 11.2). Its body shape is well adapted to gliding over the seabed, where it descends to push into the soft sediment and scoop out shelly prey items such as infaunal clams. The ray's mouth is armored above and below with an array of bar-shaped palatal teeth that function more as a crushing apparatus than as biting teeth (Fig. 11.6). The smooth outer surfaces of the opposing plates are slightly curved, and fit together as a roller-like mechanism to crush the mollusk shells on which the animal feeds. The rough inner surface on the plates represents the boney structure fused against the upper and lower parts of the mouth. The ray grinds clamshells and other shelled mollusks into small pieces in order to access the fleshy inner tissues of the prey. Immature rays may be taken by sea lions and orcas are known to attack and kill manta rays.

Sharks possess a sleek and hydrodynamic profile with a powerful tail unlike other fish, which functions to keep the heavy head in an upright position as the animal swims forward. Most fish exhibit tail fins that are Y-shaped in outline. The tail end of a shark is not forked in this way, but exhibits only an asymmetrical upper extension (called a **heterocercal tail**). At the front end, the mouth is armed with rows of backward pointing teeth. In many species, these teeth are serrated like steak knifes perfectly adapted for slicing flesh. Sharks are able to grow new teeth throughout their life. They have few natural predators aside from orcas.

Reptiles. Five out of seven of the world's sea turtles inhabit the Gulf of California at least on a seasonal basis (Seminoff, 2010). These include the green turtle (*Chelonia mydas*), hawksbill turtle (*Eretmochelys imbricata*), leatherback turtle (*Dermochelys coriacea*), loggerhead turtle (*Caretta caretta*), and olive Ridley turtle (*Lepidochelys olivacea*). In the Gulf of California, the green turtle is regarded as a regional subspecies known as the black turtle. It is the most abundant of the sea turtles frequenting the gulf, and the most commonly found in coastal waters. It also is the largest sea turtle, with adults weighing as much as 154 kg (340 lbs). The hawksbill also frequents coastal waters, but is more often seen in the southern part of the gulf. The others are most commonly observed in an open marine setting. Green turtles are mainly plant eaters, feeding on green algae and sea grass. They also consume jellyfish and some marine invertebrates such as the sea hares, which makes them omnivores. Sea turtle eggs and baby sea turtles are part of the natural food web, providing sustenance through many links in the food chain.

Marine mammals. The long-beaked common dolphin (*Delphinus capensis*) is the most abundant cetacean mammal living on a permanent basis in the gulf region, with pods (herds) that may number in the hundreds of individuals. Their body shape is sleek, gray in color with a prominent dorsal fin

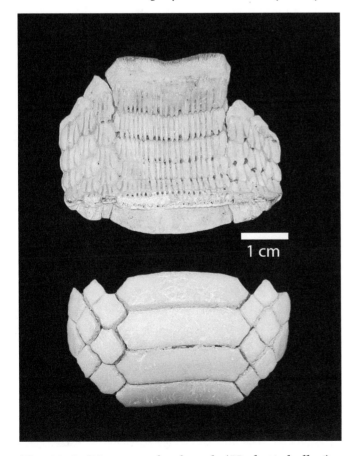

Fig. 11.6. Stingray palatal teeth (*Urobatis halleri*) from Isla Coronados.

and white belly. Adults may reach a body length of 2 to 2.5 m and live to be about 40 years of age. The dolphin's diet consists mainly of anchovies, sardines, and squid. Dolphins may participate in co-operative hunting ventures with partners who collectively chase fish into dense swarms, and then take turns picking off individuals to feed on.

After the dolphin, the California sea lion (*Z. californianus*) is one of the more common permanent residents to be found in the gulf. These pinipeds ("feather feet") have strong flippers that make them acrobatic swimmers. The rear flippers may be folded forward to give added agility on land. The adult male sea lion achieves a body length up to 2.2 m and body mass weighting as much as 390 kg. Sea lions live in rookeries on rocky promontories, where a dominant male rules over several females. Active colonies are found, for example, on the east side of Isla Coronados, Islas las Galeras (Plate 4, color section), and the southeast end of Isla Cerralvo. Sea lions are voracious eaters, consuming a wide range of fish but especially sardines and invertebrates such as squids. Other than the orca and large sharks, they have few natural enemies. A healthy sea lion may live for two or three decades.

The baleen whales include the blue whale (*B. musculus*) and the humpback whale (*M. novaeangliae*). The adult blue whale may reach a length of 30 m and attain a weight of up to 200 tons. It preys on zooplankton by lunge feeding during deep dives, when enormous numbers of krill and copepods are captured within the mouth. The front part of the mouth features a curtain of fibrous material (baleen plates) that prevent the zooplankton from escaping as seawater is expelled from the mouth by pressure from the throat and tongue. A healthy blue whale may reach an age of about 80 years. The adult humpback whale is much smaller with a maximum length of 12 to 16 m and body weight of up to 36,000 kg. Its stocky body, distinctive back hump, and long pectoral fins combine to make this species readily recognizable. The fluked tail is typically raised high above the water surface at the start of a dive sequence (Fig. 11.7). Baleen plates are located along each side of the mouth and the focus of feed-

Fig. 11.7. Tail flukes, humpback whale (*Megaptera novaeangliae*) near Isla San Ildefonso.

ing is on krill and small fish. A healthy humpback whale is estimated to reach an age of 100 years.

Birds. The brown pelican (*Pelecanus occidentalis*) maintains a nearly ubiquitous presence along the shores and islands throughout the Gulf of California. An adult bird has a wingspan of 2.5 m and may weigh as much as 5.5 kg. They are gregarious birds that live in flocks including both sexes throughout the year. In search of prey, pelicans are elegant flyers able to glide low and steady over the water surface with only few wing-beats. When fish are spotted, the pelican makes a steep dive that typically plunges the bird below the water surface for a brief time. Fish are scooped up in a large bill equipped with a gular pouch that allows water to drain out of the mouth. Herring and minnows are among the favored prey items (Fig. 11.2).

The yellow-footed gull (*Larus livens*) is a common, non-migratory bird living in the Gulf of California. The adult gull has a wingspan of up to 160 cm and weighs as much as 1,500 gm. In habit, the yellow-footed gull is a scavenger feeding on carcasses of marine mammals, but also a forager feeding on small fish and marine invertebrates. Gulls have been observed to pick up small, live clams with valves closed and drop them on rocks in order to open the shells for access to the meat within. Two kinds of grebes, Clark's grebe (*Aechmorphorus*

114

clarkii) and the western grebe (*A. occidentalis*) are frequent visitors to the Gulf of California. Both have wingspans of about 60 cm, although the western grebe is a slightly larger bird. The Clark's grebe is distinguished by its bright yellow bill in contrast to the more olive-yellow beak of the western grebe. These gregarious birds are excellent divers that feed on small fish. Flocks are commonly observed floating together on the sea surface. When one initiates a dive, the others quickly follow suit and disappear under water for a brief time only to resurface as a group once again.

Coastal Flotsam of Vertebrate Life

A survey based on the amount of vertebrate bones washed onto the beaches in the Colorado River delta area (Liebig et al., 2003) shows that the remains of several species out of 18 marine mammals known to live in the region leave a potential fossil record in beach sands and gravels. The following taxa recorded as represented by beach flotsam include the California sea lion (*Z. californianus*), common dolphin (*D. delphis*), bottlenose dolphin (*Tursiops truncatus*), vaquita (*Phocoena sinus)*, false killer whale (*Pseudorca crassidens*), pygmy sperm whale (*Kogia breviceps*), sperm whale (*Physeter macrocephalus*), and a likely beaked whale (*Mesoplodon* sp.). Three whole carcasses and 470 extraneous bones were identified from 112 localities along a single stretch of shoreline extended over four kilometers. Surprisingly, the flotsam included 28 marine mammal skulls. Such a survey underscores the fact that a beach walker should be prepared to find more than pretty shells washed onto the shore. More significant, such a survey draws a strong link to geological beach deposits as places that must be routinely checked for vertebrate fossils.

Record of Vertebrate Fossils

The record of fossil bones from marine vertebrates preserved in sedimentary deposits around the Gulf of California remains slim but taxonomically rich with some truly exceptional discoveries made in recent years. There is every expectation new discoveries will continue to be made due to exploration of more remote localities by those naturalists who enjoy coastal hiking. The accompanying map (Fig. 11.8) shows the distribution of vertebrate fossil localities divided between Pleistocene and Pliocene sites numbered from north to south. Discoveries also made outside the Gulf of California on the Pacific shores of the Baja California peninsula help to fill out the picture of marine vertebrate diversity based on even older and more extensive Miocene and Oligocene deposits with related ancestral stocks of the animals that later reached the Gulf of California after its tectonic opening (Barnes, 1998). However, this review is focused only on the distinct local faunas within the Gulf of California.

Pleistocene. Deposits of this age with vertebrate remains are under-reported in the scientific literature. According to the summary of Barnes (1998), an unnamed formation on Isla Tiburón (site △ site #1) yielded a single toe bone from a sea lion (*Z. californianus*). The reference is notable both because Pleistocene material is rare and because it represents the northern most locality presently known for fossil vertebrates.

Pliocene. Basin deposits at San Francisquito (site □ #1) feature fossil shark teeth of the kind similar to the great white shark (*Carcharodon* sp.), as illustrated by Johnson (2014). A more systematic survey for vertebrate remains is needed at this locality.

The most important locality for marine vertebrates in terms of diversity is at Loma del Tirabuzón (Applegate and Espinosa-Arrubarrena, 1981). Located on the coast 5 km north of Santa Rosalía (site □ #2), the spot also is known as Corkscrew Hill due to preservation of spiral-shaped burrows made by ten-legged crustaceans (Barnes, 1998). The Tirabuzón fauna includes the remains of sharks, rays, bony fish (Applegate, 1978; Applegate and Espinosa-Arrubarrena, 1981), as well as sea turtles, sea lions, and whales. According to Barnes (1998), the fauna also features bones belonging to the common dolphin, a porpoise similar to the modern Pacific harbor porpoise (*Phocoena phocoena*), a pygmy sperm whale similar to a living

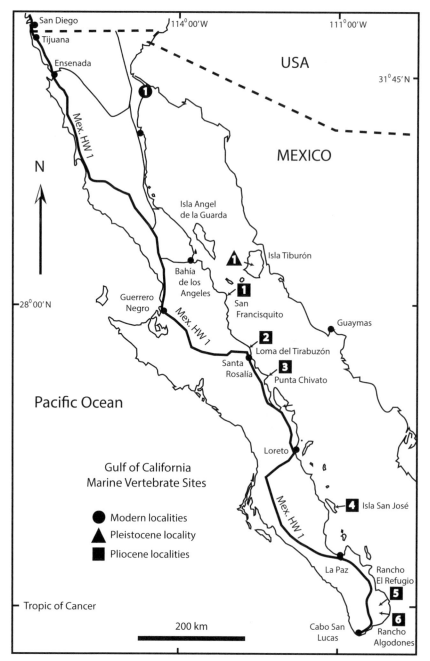

Fig. 11.8. Map showing modern and fossil vertebrate deposits in the Gulf of California.

species (*Kogia* sp.), an extinct pygmy sperm whale (*Scaphokogia* cf. *cochlearis*), and the tooth of a true sperm whale.

Pliocene strata at Punta Chivato close to the community trash site (site □#3) are known by collectors to contain abundant shark teeth, as yet not adequately studied (Johnson, 2002). From Isla San José (site □ #4), comes a newly recognized Pliocene dolphin (*Protoglobicephala mexicana*) regarded as extinct based on well-preserved skull bones (Aguirre-Fernández et al., 2009).

Pliocene sediments from the Trinidad Formation at Rancho El Refugio 17 km southeast of Santiago (site □ # #5) provide information on another diverse vertebrate assemblage including sharks, fish, and marine mammals such as an extinct walrus-like animal (*Aivukus cedrosensis*). This is the most southern locality to yield fossil pinnipeds in the Pacific realm (Barnes, 1998). Similar deposits at nearby Rancho Algodones (site □ #6) also are attributed to the Refugio Formation, and have yielded parts of whale skulls and the large enamel-covered crowns identified as belonging to

116

a sperm whale (*Scaldicetus* sp.). Likewise coming from the same Rancho Algodones area (Fierstine et al., 2001), one of the most spectacular fossil discoveries in recent years is represented by the fossil skull and bill fragments of a blue marlin (*Makaira nigricans*).

In direct comparison with Pliocene and Pleistocene deposits containing abundant marine invertebrates, the existing record of vertebrate remains in the Gulf of California is sparse. However, the trail of important discoveries made during the last 35 years is more than sufficient to underscore the fact that a complex food web was already in existence soon after the major flooding of the Gulf of California approximately five million years ago. Marine mammals including large whales frequented the paleo-gulf early on in its history and added to its rich biodiversity. Fossil evidence as circumspect as the broken fragments of whale ribs from Pliocene shore deposits near Punta Chivato (Plate 4, color section) lend mute testimony to this fact.

Chapter 12

Conservation and Sustainable Development

Introduction

The model food pyramid and food web for the Gulf of California from the preceding chapter (review Figs. 11.1 and 11.2) exclude any trace of influence from the ecologically dominant species represented by humans (*Homo sapiens*). Folk have lived around the Gulf of California for thousands of years, as witnessed by rock art and by kitchen middens left behind by relatively small populations that moved along the shores and inland living as hunter-gatherers. In 1684, settlers from mainland Mexico who attempted to establish a mission at San Bruno near present-day Loreto witnessed an assembly of some 2,500 Indian inhabitants gathered for ceremonial rites (O'Neil and O'Neil, 2001). Such a sizable population would not have been a permanent presence at a single coastal locality due to the limitations of food and drinking water. Indeed, Europeans quickly abandoned the San Bruno settlement due to an inadequate supply of drinking water.

As many as a dozen linguistic groups occupied the Baja California peninsula, but three tribal groups dominated much of the region. The Cochimi peoples were dispersed over the largest area stretching from about present-day San Felipe southward below the Concepción peninsula, but present on both the Pacific and gulf sides of Baja California. Also on both coasts, the Guaycura peoples occupied a smaller region that included Isla del Carmen as well as the Sierra de la Giganta and the Llano de Magdalena. The Pericú peoples inhabited the entire southern Cape Region, but also frequented the gulf islands of San José, Espiritu Santo, and Cerralvo. On the opposite mainland shores, native peoples such as the Seri, Yaqui, and Mayo populated extensive areas fronting the Gulf of California. Among them, only the Seri retain a significant presence now largely limited to Isla Tiburón in Sonora. Extensive ethnographic research on the Seri,

for example, reveals how a wide array of shellfish harvested from the Gulf of California continues to form an important part of their diet and culture (Marlett, 2014). Clearly, the population size of all regional indigenous groups was small enough to insure that marine ecosystems in the Gulf of California remained viable over time to supply a lasting and renewable source of protein.

Modern times have brought greater economic pressures to bear on the region, eroding the vitality of gulf ecosystems to the extent they no longer represent the same pristine collection of components as they once did 475 years ago when the mariner Francisco de Ulloa sailed under orders from Hernán Cortés to explore Mexico's western frontiers. Today, the rich fisheries of the Gulf of California are estimated to yield about half of Mexico's total production for the entire country (Brusca, 2010). In excess of 500,000 tons of seafood are harvested each year from the Gulf of California, a quantity that fails to account for the wasted by-catch discarded by the trawler fleet long before desired products reach market. On a global scale, the threat of exhausted natural resources has prompted ecologists to consider the sustainability of **ecosystem services**, broadly divided between direct and indirect benefits to society.

Direct services include the provision of food, clean water, and fresh air, whereas indirect services entail non-material advantages that satisfy aesthetic, spiritual, and recreational values held by those who live in a particular region and those who visit as eco-tourists. Should growing economic pressures require greater oversight, it is argued that the global monitoring of coastal ecosystems must be performed in a more efficient way to insure their continued health. That sentiment clearly applies to the Gulf of California (Santamaria-del-Angel et al., 2015).

The aim of this concluding chapter is to review the history of significant steps taken at the Mexican federal level to protect ecosystem services in the Gulf of California. More so, it is the goal of this summary to offer an outline for the enhancement of indirect ecosystem services through sustainable community development that builds on a fusion between living coastal ecosystems and the **geoheritage** represented by past ecosystems in the Gulf of California preserved in the region's remarkable rock record reaching back five million years in geologic time. The gulf region includes a wealth of outstanding **geosites** that deserve the protection and admiration both of local residents and visiting eco-tourists. That the region truly possesses such an untapped treasure of geosites and the potential for sustainable economic development based on expanded eco-tourism has only begun to be recognized on the world stage (Bruno et al., 2014).

Growth of a Conservation Ethic

Outside visitors have long been attracted to the Gulf of California to study the region's ecological and geological resources, sometimes with the express support of Mexican authorities but just as often without any formal agreement. The Italian naturalist Federico Craveri (1815-1890) was an early explorer, who visited the Gulf of California in 1856 as a government agent sent to assess the region's guano deposits as a source of fertilizer. Craveri's extensive journals (all in Italian) record observations on the full range of nature from the relationship between geography and geology to descriptions of birds, fish, and marine invertebrates. Craveri's murrelet (*Synthliboramphus craveri*) is a bird that nests in the upper gulf region, and it was new to science at the time of his visit. Previously under-estimated, the Italian's contributions are being re-evaluated and a translation of his work is in progress (personal communication, Thomas Bowen, 2015). Another visitor from afar was the Hungarian naturalist, John Xántus (1825-1894), who manned a tidal-gauge station established by the U.S. Coast and Geodetic Survey at Cabo San Lucas in 1859. Xántus also is credited with discovering species new to science during his time

in the Cape Region of Baja California. Extensive natural history collections made by Xántus came to the Smithsonian Institution in Washington, D.C., where specialists formally described many of his fish specimens new to science. Craveri eventually returned to Italy, where he became a professor at Turin University. Xántus returned to Hungary, where he became director of the Budapest Zoological Garden and worked as curator of ethnography in the Hungarian National Museum.

Subsequent visitors had a profound impact on natural history studies in the Gulf of California. March 2015 marked the 75th anniversary of the much-celebrated voyage in 1940 to the Gulf of California by marine biologist Ed Ricketts (1897-1948) and writer John Steinbeck (1902-1968) aboard the *Western Flyer*. Both are honored with bronze busts of their likenesses in Monterey, California (Figs. 12.1 and 12.2). An account edited as *The Log from the Sea of Cortez* (Steinbeck, 1951) has become a cult classic much admired for the pair's holistic view of nature expressed well before the word ecology achieved the common usage it now enjoys. By 1939, Ricketts' textbook *Between Pacific Tides* was

Fig. 12.1. Monument to Edward F. Ricketts (1897-1948).

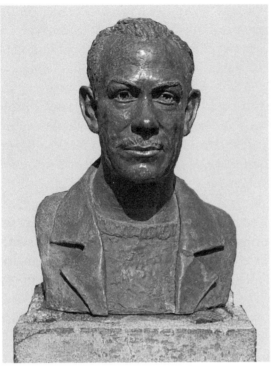

Fig. 12.2. Monument to John Steinbeck (1902-1968).

published and Steinbeck found success as the author of *The Grapes of Wrath*. With his earnings as a novelist, Steinbeck chartered the 76-foot commercial purse seiner out of Monterey, California, for a six-week expedition to the Gulf of California. At the site of the biologist's early laboratory in Pacific Grove, a marker reads (in part): "Ricketts' studies of marine intertidal life became benchmarks in ecological science. Friend John Steinbeck immortalized him as Doc in the novels *Cannery Row* and *Sweet Thursday*."

October 2015 also marked the 75th anniversary of the voyage by the 104-foot research schooner *E.W. Scripps* to the Sea of Cortéz in 1940. Under the command of Captain E.D. Hammond, the expedition lasted 78 days, with 65 days devoted to research within the Gulf of California. The Scripps Oceanographic Institute and Geological Society of America co-sponsored of the excursion, which logged 4,600 nautical miles within gulf waters. Among the scientific crew were geologist Charles A. Anderson (1902-1990), paleontologist J. Wyatt Durham (1907-1996), and oceanographer Francis P. Shepard (1897-1985). Francisco Diaz-Salcido accompanied the trip as a representative of the Mexican government. Expedition results became available in 1950 as a multi-part report issued by the Geological Society of America.

The GSA volume features a set of bathymetric charts compiled by Shepard, including a master chart for the whole region. For the first time, it was possible to visualize the configuration of several deep-water basins that partition the Gulf of California. Today with input by geophysicists, these segmented basins are understood as part of the gulf's tectonic framework (see Chapter 1, Fig. 1.1). Geologist Anderson was the first to map rock formations on several of the islands in the Gulf of California, including Isla del Carmen where massive conglomerate beds are now regarded as part of a lost river system formerly connected to the Sierra de la Giganta (see Chapter 9). Paleontologist Durham described a rich trove of invertebrate fossils from arroyos Arce and Gua, a short distance north of Loreto through one of the most complete stratigraphic sections found in the gulf region of Baja California. The 1940 cruise of the *E.W. Scripps* is not lost to history, because it set a high standard for scientific research valued to this day.

With chapters on ecology and paleoecology paired for matching environments, the *Atlas of Coastal Ecosystems in the Western Gulf of California* (Johnson and Ledesma-Vázquez, 2009) brought together a team of nine Mexican specialists and seven from the U.S.A. to provide coverage augmented by a comprehensive set of satellite images for the entire coastal zone and related islands flanking the gulf shores of the Baja California peninsula. The volume's contents echo contributions of the 1940 expeditions by the *Western Flyer* and the *E.W. Scripps* by placing an emphasis on connections between today's Gulf of California and its geological past. A robust spirit of international co-operation also was fostered. A key advance subscribed to by this project was the proposal that various coastal ecosystems in the Gulf of California conform to geographic boundaries influenced by large-scale patterns of wind and wave circulation (see inside back cover). Such patterns have roots in the geologic past that have persisted for millions of years.

The Gulf of California has yet more secrets waiting to be discovered. Regarding the same invertebrate fauna that brought the *Western Flyer* expedition in 1940, it is estimated that the 4,900 species now described and formally named by marine biologists represent only 70% of the potential taxa (Brusca and Hendrickx, 2010). Fresh discoveries are anticipated in all areas of biology and paleontology, if only because large areas represented by coastal ecosystems within the Gulf of California remain yet to be studied in close detail. That which deserves and requires conservation can only be protected based on direct knowledge of what actually exists in an unusually vast and complicated setting that supports multiple interlocking ecosystems. Acknowledgement that the Gulf of California and adjoining Baja California peninsula still embodies one of the great wilderness areas intact on planet Earth demands ongoing devotion to a conservation ethic for the mutual benefit of nature, the people who live in the region, and those who wish to visit.

Mexico's System of Bioreserves

In Mexico, Natural Protected Areas (ANP for Áreas Naturales Protegida) occupy 8.7% of the country's territory, covering about 1,700,000 km² (Olvera et al., 2004). Those lands and marine zones under conservation include biosphere reserves, national parks, and other protected areas for flora and fauna (Bourillón and Torre, 2012) as established under Presidential Degree managed by the Secretary of the Environment and Natural Resources (SEMARNAT) through the National Commission of Protected Natural Areas (CONANP). Protected areas include, for example, marine zones for conservation of the Vaquita porpoise (*Phocoena sinus*) in the upper Gulf of California through the Instituto Nacional de Ecología (INE). The largest of the protected areas on the Baja California peninsula is El Vizcaino Biosphere Reserve (143,600 km²), formalized on November 30, 1988 (inscribed by the Diario Oficial de la Federación for that year). The reserve features important coastal zones both on the Pacific Ocean side of the peninsula (including the Laguna Ojo de Liebre with its seasonal whale population) and the gulf coast. The extent of natu-

ral protected areas situated on the Baja California peninsula and its associated islands amounts to the greatest share of lands under conservation in all of Mexico, roughly 19% of the nation's total reserves.

Federal efforts began in Baja California in 1938 when President Lazaro Cardenas declared a protected zone around the City and Port of La Paz. Thereafter, island lands within the Gulf of California became the focus for greater attention. Isla Tiburón gained protected status in 1963 (listed for that year in the Diario Oficial de la Federación), and only a year later Isla Rasa gained the same status (listed in the Diario Oficial de la Federación for 1964). Subsequently, a decree was issued in August 1978 a to establish a reserve and refuge for migratory birds and the wildlife living on all islands in the Gulf of California. This enormous zone that encompasses a composite island territory of 2,977 km² was formally declared an "Area of Protection of the Flora and Fauna" in the Gulf of California on June 7, 2000. Five years later on July 14, 2006, the island biosphere reserve system achieved the commendation of the United Nations Educational, Scientific, and Cultural Organization (UNESCO) with the official registration of the system on its list of sites having "outstanding universal value which deserves protection for the benefit of all humanity" (Fig. 12.3). In practice, management of this area is complex as it comes under three regional directorates and prioritization of resources directed to conservation efforts on each of the separate islands is problematic.

Fig. 12.3. UNESCO commemorative plaque (La Paz).

Cabo Pulmo National Park

Cabo Pulmo National Park (Parque Nacional Cabo Pulmo) entails the only extensive coral reef in the Gulf of California and, as such, protects a special habitat with ecological processes, biological communities, and physiographic features all its own. The coral reef at Cabo Pulmo is the most northern reef found in the Eastern Pacific and one of the largest. Located about 96 km north of Cabo San Lucas on the gulf coast and encompassing a marine zone more than 70 km^2 in size, the park was granted status as an ANP on June 6, 1995 with the specific goal of protecting the coral-reef ecosystem and associated biodiversity unique to the Gulf of California. In particular, the park reef represents the most extensive coral coverage in the Gulf of California, inhabited by 11 of the 14 species of hermatypic corals reported for the region. The fish fauna also is considered important, because 226 of the 875 fish species reported for the Gulf (26%) are typically present within park waters. With regard to marine mammals, it is possible to observe a small colony of sea lions, as well as three kinds of dolphins. In addition, the park commonly hosts five of the world's seven sea turtles. These come under the category of endangered species in Mexico. During the winter season, several kinds of whales are known to frequent the park.

Loreto Bay National Park

Loreto Bay National Park (Parque Nacional Bahía de Loreto) earned ANP status on July 19, 1996, with the specific aim to protect ecosystems and associated biodiversity within the bay, as well as to promote the economic development of local communities settled in the area. This national park is located along the gulf shores of the Loreto municipality. Marine zones under protection incorporate islas Coronados, Carmen, Danzante, Monserrat and Santa Catalina (or Catalana) and various islets (Fig. 12.4). In total, the park covers an area of 206 km^2 of which the islands and islets comprise nearly

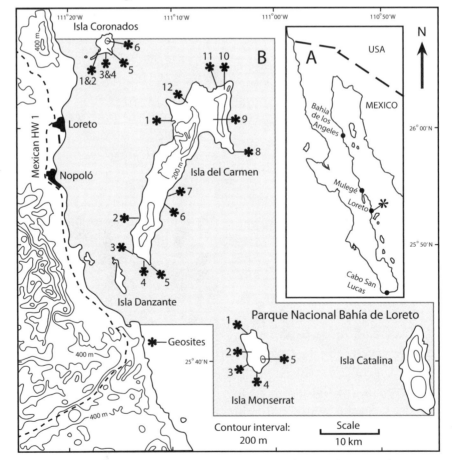

Fig. 12.4. Parque Nacional Bahía de Loreto boundaries showing key geosites.

12%. The Loreto Bay National Park features several coastal, insular and marine habitats, and serves as a refuge for a large number of marine and terrestrial species.

The park flora is represented by 161 species of microalgae and more than 600 species of terrestrial plants registered to the Islands. Fish populations stand out for their abundance with over 260 species recorded for the park, among which the ornamental fish belonging to the rocky reef zone are highlighted. In addition, the park commonly hosts five of the seven of the world's sea turtle genera. Marine mammals also are well represented with registration of as many as 30 species. They commonly include sea lions, dolphins, and as many as six different kinds of whales. In all, 75% of marine mammals found off Mexican shores can be observed within the marine park. As regards birds, about 228 species of resident and migratory birds are recorded. On the other hand, reptiles are more representative of the island terrestrial species, whereas mammals (including rodents) are the least diverse. The island terrestrial fauna is notable for a high degree of endemism.

Role of Geoparks

The geopark concept evolved as a program for the protection of areas with special geological features. UNESCO adopted the program in 2001, giving the following definition: "A global geopark is a unified area with a geological heritage of international significance." Successful geoparks exhibit links to geoheritage in order to promote awareness of major issues facing society in the context of our dynamic planet. Many promote awareness of geological hazards, including volcanoes, earthquakes and tsunamis. Some help to codify disaster mitigation strategies among local communities. Geoparks hold the physical records of past climate change and serve to instruct visitors on current climate change. They also help to focus attention on the best approach for utilization of renewable energy and employment high standards for green tourism. Above all, geoparks foster sustainable economic development in adjoining communities.

Geoparks do not necessarily carry a legislative mandate, although key geosites within a geopark may be protected under appropriate legislation at the national or state level. The Global Network of National Geoparks is a voluntary network supported by UNESCO, whose members are committed to the exchange ideas about management and sometimes join together on projects of common interest. As described by Crofts and Gordon (2015), **geodiversity** expresses the natural range of topics that include features of geological (rocks, minerals, fossils), geomorphological (landforms, topography, physical processes), as well as hydrological and agronomical importance.

Earth's geodiversity has ramifications that contribute directly to most of the ecosystem services recognized in the Millennium Ecosystem Assessment (2005). It provides the foundation for plants, animals and humans, and it forms a vital link between people, nature, landscapes and cultural heritage. Above all, it provides assets for outdoor recreation and enjoyment of the natural world. An understanding of how the Earth works is key to all levels of managerial planning for lands, rivers, and coastal zones. This is particularly critical during at a time of uncertainty about the full effects of climate change and sea-level rise. Hence, it is appropriate that the concepts of geodiversity and geoheritage entailed in geoparks become well integrated into the management of protected areas and recognized as equally important for the sustainability of biodiversity as part of an ecosystem approach that values the integrity of abiotic and biotic processes.

Confluence of Natural Protected Areas and Geoparks

Natural Protected Areas (ANPs) are the political tool of choice commonly applied as the most effective means for the conservation of natural resources. However, the planning and management that goes into ANPs typically becomes focused mainly on the preservation of biodiversity. As a consequence, political action is readily reduced to a mono-vision of nature entirely separated from human societal issues. Procedural limitations associated with ANPs often entail restrictions on land use without the consensus of local populations, thereby provoking conflicts of interest. Nonetheless, the achievements in conservation achieved by ANPs at the global level are undeniable. In Latin America, however, this form of conservation stands at only about 15% coverage of the entire territory.

In contrast, the geopark concept seeks to provide alternatives to land use that may already possess conservation structures (like an ANP). Geoparks and ANPs are not mutually exclusive entities, but can be regarded as potentially complementary in terms of shared environmental, economic and social agendas (Fig. 12.5). The geopark structure has a basis in community participation coming from the bottom up and promotes local efforts to blend elements of science and education with eco-tourism and community development. It affords the opportunity to involve communities in management and decision-making over land-use planning and nature conservation in their own back yard, hence fostering a sense of local ownership and belonging valued as a vital socio-environmental asset. The basic requirements for development of a geopark are twofold: 1) that it exhibit aspects of geological heritage having international importance suitable for development of educational and outreach strategies, and 2) that it generates economic activities in the communities embedded

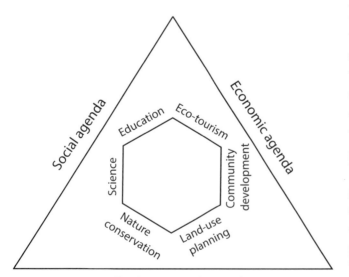

Fig. 12.5. Tri-part agendas for geoparks.

within the territory, mainly through eco-tourism and other innovative activities.

Evaluation of Geosite Clusters

A viable geopark depends on a cluster of geosites that function together to represent the geoheritage of a particular district at a high level of integration. As a working unit, a geosite is a geological object or part of the geological setting exposed at the land surface, and hence accessible for visitation and study (Ruban, 2010; Bruno et al., 2014). Because it is not possible to protect all geological phenomena in any given district, choices must be made to select those geosites grouped together in such a way to best meet standards for scientific and educational use in the context of site conservation. Criteria such as the degree of representation and rarity have a bearing on scientific relevance, whereas accessibility and visibility have a direct impact on educational utility.

Geosites may feature elements that are small in scale (for example, a single rock outcrop) or demonstrate larger patterns that show a geological transition or connectivity in a broader paleogeographic context (Gonggrijp, 2000; Bruno et al., 2014). Geosite assessment is subject to the following criteria (Wimbledon et al., 2000): 1) significance for understanding geological or geomorphological processes as well as geological evolution (abiotic and biotic), 2) completeness with regard to the phenomena present, 3) degree of uniqueness shown by features in time and space (in reference to physical scale and span of geological time occupied), 4) overall quality of physical materials, 5) degree to which clustered geosites represent broad thematic patterns iconic for the surrounding region, and 6) the extent to which the site already has been studied and the availability of literature regarding the site. In addition, quite another aspect strictly independent of geological factors is of critical importance with regard to the development of a geopark formed around a fitting cluster of geosites. To what degree can societal and political alliances be formed to assure the adequate conservation of lands around a potential geopark?

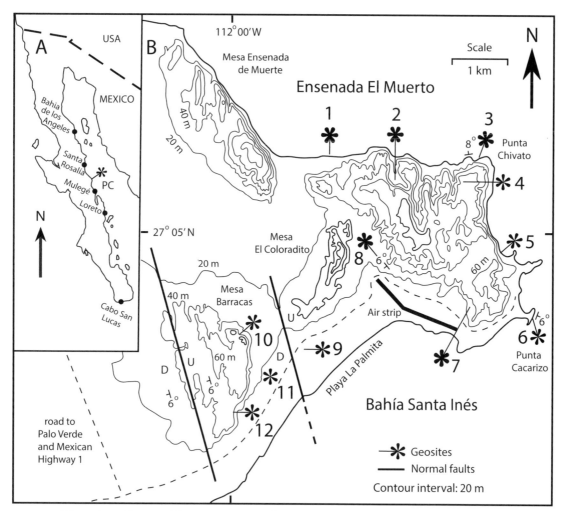

Fig. 12.6. Punta Chivato region showing geosite clusters.

Punta Chivato Geosites

Located between Santa Rosalía and Mulegé (Fig. 12.6A), the Punta Chivato region features a wide promontory (or *atravesada*) that projects crossways into the Gulf of California for several kilometers at elevations up to 100 m above sea level (Fig. 12.6 B). The area has been the focus of intensive studies on Pliocene and Pleistocene geology with a strong focus on paleogeography (Simian and Johnson, 1997; Libbey and Johnson, 1997; Johnson, 2002a; and Johnson, 2002b). In particular, the district features a group of mesas that rise above the surrounding countryside and represent separate islands belonging to a former archipelago dating back as much as five million years. The paleoislands, as well as prominent geomorphological features exposed along the present coastline, contain a cluster of a dozen key geosites (Fig. 12.6B). Shores that border on Ensenada El Muerto have a northern exposure

that is often subject to strong wind and sea swells, whereas southern shores provide leeward shelter to the calm waters of Bahía Santa Inés.

Punta Chivato geosites relate to patterns of wind and waves that remained persistent over geologic time. Moreover, they also reflect a history of tectonic change in concert with changing patterns of global sea level. Hence, they offer a combination of themes pertinent to the tectonic, geologic, and environmental evolution of the Gulf of California. Some Punta Chivato geosites are featured in the preceding chapters of this volume, as for example the Pleistocene phantom island at Playa La Palmita (Fig. 2.11), the sea cliffs showing the onshore migration of Pliocene conglomerate and limestone beds across a volcanic paleoshore (Fig. 2.13), the modern shell beach and associated sand dunes on the Ensenada El Muerto (Fig. 6.5), and fossil stromatolites from a Pleistocene lagoon now elevated

125

on the promontory nearby Punta Chivato (Figs. 8.8 and 8.9). Hiking paths that connect these and many other sites are laid out in a guidebook specifically dedicated to the region (Johnson 2002a). The following list (Table 12.1) provides notes and reference citations to the dozen Punta Chivato geosites identified on the map in Figure 12.6B.

Geosite	Description	Reference
1	Modern shell beach	Johnson (2002, p. 103)
2	Pliocene sand dunes	Johnson (2002, p. 111)
3	Pliocene limestone ramp (north shore)	Johnson (2002, p. 80-81)
4	Pleistocene stromatolite lagoon	Backus & Johnson (2014)
5	Pliocene coastal onlap and unconformity	Johnson (2002, p. 53)
6	Punta Cacarizo unconformity (east shore)	Johnson (2002, p. 43)
7	Pleistocene coral reef (south shore)	Johnson (2002, p. 67)
8	Pliocene oyster bank (leeward ramp)	Johnson (2002, p. 96-97)
9	Pleistocene Phantom Island	Johnson (2002, p. 33-34)
10	Pliocene oyster bank	Johnson (2002, p. 123)
11	Pleistocene sand dune	Johnson (2002, p. 145)
12	Pliocene submarine slump	Johnson (2002, p. 134)

Table 12.1. Geosites in the Punta Chivato region.

The Punta Chivato region is an ideal setting that satisfies criteria for evaluation of a geopark as reviewed above (Gonggrijp, 2000; Bruno et al., 2014). With an existing guidebook to the area (Johnson, 2002a), it offers a set of *de facto* geosites. Unfortunately, the region's complicated tangle of land ownership and fence lines are a major detraction that diminish the candidacy of Punta Chivato as a potential geopark. Some localities available as geosites in 2002 are now lost to road development and the ongoing sprawl of vacation homes. However, this drawback should not dissuade eco-tourists from visiting and exploring the area.

Loreto Area Geosites

With protection of the Loreto Bay National Park as a legitimate ANP (Figs. 12.4A and B), several islands under park jurisdiction offer exceptional potential for development of a successful geopark. In particular, three islands possess the same qualities found in rock outcrops and other large-scale features denoting geosites in the Punta Chivato region.

Isla Coronados. The smallest but most accessible park island is Isla Coronados, situated north of Loreto and covering an area of 7.59 km^2 (Carreño and Helenes, 2002). The island has superb sandy beaches, a Pleistocene stratovolcano, and a Pleistocene coral reef preserved in a former lagoon on the island's south side. These features are the subject of detailed scientific studies (Bigioggero et al., 1988; Johnson et al., 2007; Ledesma-Vázquez et al., 2007; and Sewell et al., 2007). The island's topography is well mapped showing existing trails. A book chapter is devoted to a half-dozen geosites, as well as several other points of interest on the island (Johnson, 2014). The following list (Table 12.2) provides notes and reference citations to the key Isla Coronoados geosites identified in Figure 12.4 B.

Geosite	Description	Reference
1	Modern rhodolith sand beach (west side)	Johnson (2014, p. 146)
2	Modern rhodolith sand dunes	Johnson (2014, p. 147)
3	Pleistocene lagoon entrance (south shore)	Johnson (2014, p. 152)
4	Pleistocene barrier over-wash deposits	Johnson (2014, p. 154)
5	Pleistocene coral reef	Johnson (2014, p. 156)
6	Pleistocene volcanic cone (island center)	Johnson (2014, p. 160)

Table 12.2. Key geosites for Isla Coronados (Parque Nacional Bahía de Loreto).

Isla del Carmen. With an area of 143 km^2, Isla del Carmen is the largest in the Loreto Bay National Park (Carreño and Helenes, 2002). Located across from Loreto (Fig. 12.4B), it is elongate in shape and separated from the peninsular mainland by the Loreto Passage with water depths that reach 460 m toward the deeper north end of channel. The island features extensive limestone formations of Pliocene and Pleistocene age that occur in places along east and west coasts, as well as the shorter north and south ends. The island has been the focus of several recent studies on limestone and conglomerate deposits (Dorsey et al., 2001; Eros et al., 2006; Johnson et al., 2009; Backus and Johnson, 2009; Johnson and Backus, 2009; and Ledesma-Vázquez et al., 2012). The main emphasis of these studies relates to the orientation of geological deposits around the island's periphery with respect to patterns of prevailing wind and sea swell, although tectonic uplift and changes in sea level are ancillary themes. In this regard, as many as a dozen key geosites on Isla del Carmen (Table 12.3) strike an especially favorable thematic comparison with those from the Punta Chivato region (consult the geographic/paleogeographic patterns from the inside, back cover).

Geosite	Description	Reference
1	Pleistocene coral reef	López-Pérez (2008, p. 506)
2	Pliocene rhodolith beds (Bahía Marquer)	Johnson et al. (2009, p. 91)
3	Marine terrace steps (southwest cliffs)	None
4	Pleistocene rhodolith cliffs	Cover, this volume
5	Modern rhodolith beach sand	Cover, this volume
6	Pliocene rhodolith beds (Arroyo Blanco)	Eros et al. (2006, p. 1153)
7	Pliocene river delta	Johnson et al. (in press)
8	Pliocene submarine fan conglomerate	Dorsey et al. (2001, p. 104)
9	Modern salt lagoon	Kirkland et al. (1966, p. 7)
10	Modern dun sand with shell debris	Backus & Johnson (2009, p. 122)
11	Pleistocene dune cliffs	Johnson & Backus (2009, p. 143)
12	Modern & Pliocene dunes (Bahía Otto)	Johnson & Backus (2009, p. 142)

Table 12.3. Key geosites for Isla del Carmen (Parque Nacional Bahía de Loreto).

Isla Monserrat. Nearly 20 km² in area (Carreño and Helenes, 2002), this island is situated in the southern part of Loreto Bay National Park (Fig. 12.4B). Detailed field studies have been devoted to this island (Ledesma-Vázquez, 2007; Johnson and Ledesma-Vázquez, 2009; and Johnson et al., 2009). The island's topography is mapped and a book chapter is devoted to the description of key geosites and other points of interest (Johnson, 2014). Monserrat features five key geosites that are extraordinary, both for exposures of Pliocene limestone formations that

have undergone tectonic uplift and for examples of geomorphic and depositional processes along present-day shores (Table 12.4).

Geosite	Description	Reference
1	Modern shell beach (north shore)	Johnson (2014, p. 180)
2	Pliocene rhodolith beds (high plateau)	Johnson (2014, p.176-78)
3	Modern wave-incised rocky platform	Johnson (2014, p. 174)
4	Pliocene pecten balls (submarine slump)	Ledesma-Vázquez et al. (2007)
5	Pliocene paleoshore-line	Johnson (2014, p. 171)

Table 12.4. Key geosites for Isla Monserrat (Parque Nacional Bahía de Loreto).

Presently, all islands other than Isla Coronados within Loreto Bay National Park are subject to restrictions that prohibit access to island interiors. Even so, most of the geosites denoted for Isla del Carmen and Isla Monserrat are readily observable from tour boats or kayaks. Islas Danzante and Catalina lack any record of sedimentary rocks, but under future study these islands are sure to provide coastal geosites that reflect notable aspects of geologic and geomorphologic processes.

Bahía de los Angeles Geosites

Although not as thoroughly studied as coastal regions in Baja California Sur, the area around Bahía de los Angeles in the northern state of Baja California holds enormous potential for the promotion of geosites and the possible development of a geopark. All islands in this district enjoy ANP status. An area approximately 45 km² at the northern end of the rugged Sierra Las Animas forms the southeast margin of Bahía de los Angeles (Fig. 12.7). Gastil et al. (1971, map C) were the first to map the geology of this district in any detail. In particular, they recognized some notable features that define a former marine channel about 2 km in length separating the Quemado headland from the rest of the sierra as a Pliocene island. Located 10 to 12 km east of the village of Bahía de los Angeles, this area includes a cluster of geosites with a focus on geologic. pale-

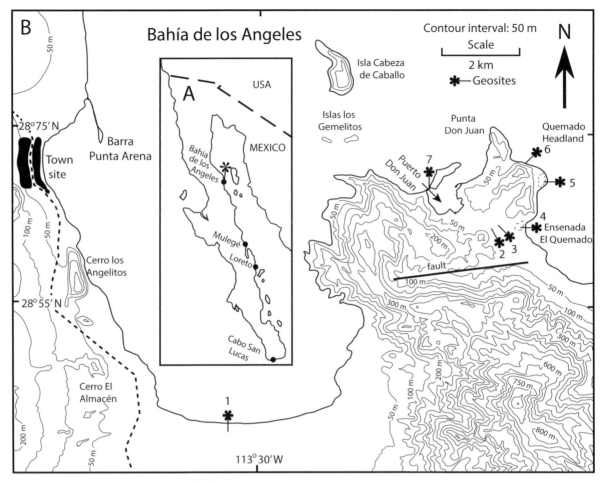

Fig. 12.7. Bahía de los Angeles area showing key geosites.

ontologic, biologic, and geomorphic features (Table 12. 4). A separate geosite located on the north-facing shore of Bahía de los Angeles also is of much interest from a biological point of view regarding wave-transported shells.

Three of these localities are treated by the present volume in Chapter 4 (geosites 1 and 3) and Chapter 8 (geosite 4). A valley trail that connects Puerto Don Juan with the Ensenada El Quemado and continues in a loop around to the Quemado headland would link together most of these sites. Additional sites of geological significance are sure to be found through future exploration of this interesting district. For example, a major east-west fault (Fig. 12.7) forms the prominent junction between uplifted igneous rocks to the north and metamorphic rocks to the south. A side trail leading to higher elevations on the Sierra Las Animas would traverse the scenic canyon that marks the fault and terminate with unparalleled views over the region.

Geosite	Description	Reference
1	Modern shell berm	This volume (p. 36)
2	Cretaceous- Pliocene unconformity	Gastil et al. (1971, map C)
3	Pliocene marine channel & bivalves	This volume (p. 39-40)
4	Modern stromatolites	This volume (p. 80)
5	Closed lagoon	None
6	Quemado headland (tonolite & basalt)	This volume (p. 39)
7	Puerto Don Juan tombolo	None

Table 12.5. Key geosites for the region east of Bahía de los Angeles.

A Closing Challenge

The fundamental premise for organization of this guidebook is that the Gulf of California's geological backdrop is equally as important and awe inspiring as the ecological play that goes on stage every day with star performances by primary players such as humpback whales, dolphins, and sea lions (Chapter 11) interacting with the panoply of life from different coastal ecosystems. A measure of exposure to the gulf's rich fossil record is enough to convince even the casual visitor that today's ecological play has had a long and highly successful run over the last several million years. The stage, itself, is one framed by tectonic regimes that brought the dynamics of sea-floor spreading in the eastern Pacific Ocean into a direct encounter with the Mexican frontier lands of the North American continent. That interface of physical geography is one that has ramifications for ongoing expansion of the Gulf of California and its eventual encroachment across the international border through the territory of the San Andreas fault and its subsidiary faults all the way to northern California.

The enormous size of the Gulf of California is such that fresh insights on its integrated ecosystems and distinct echoes from related systems belonging to the geologic past are certain to make an impression on those who take the time to explore almost any random stretch of the enclosing coastline or its associated island shores. As point and counter-point, the gulf's past and present ecosystems provide a world-class setting rich in geodiversity through which to investigate the intersection of geography, geology, and ecology across time. The challenge put forward to the individual reader is to go out to the world of nature and to remain open to the possibility of discovery. Those who call the towns and villages scattered around the Gulf of California home can take tremendous pride in the wealth of nature found close by. They are entitled to enjoy the sustainable benefits of eco-tourism as visitors come to the region to learn about and enjoy all it has to offer. In consideration of geosites and potential geoparks reflective of this history, the Gulf of California ranks second to none and deserves all the best the future has to offer.

Bibliography

Adey, P.J., and Macintyre, I.G. 1973. Crustose coralline algae: a re-evaluation in the geological sciences. Geological Society of America Bulletin 84: 883–904.

Aguirre-Fernández, G., Barnes, L.G., Aranda-Mantreca, F.J., Fernández-Rivera, J.R. 2009. *Protoglobicephala mexicana*, a new genus and species of Pliocene dolphin (Cetacea; Odontoceti; Delphinidae) from the Gulf of California, Mexico. Boletín de la Sociedad Geológica Mexicana 61: 245–265.

Allwood, A.C., Walter, M.R., Kamber, B.S., Marshall, C.P., Burch, I.W. 2006. Stromatolite reef from the Early Archean Era of Australia. Nature 441: 714–718.

Amado-Filho, G., Moura, R.L., Bastos, A.C., Salgado, L.T., Sumida, P.Y., Guth, A.Z., Francini-Filho, R.B., Pereira-Filho, G.H., Abrantes, D.P., Brasileiro, P.S., Bahia, R.G., Lean, R.N., Laufman, L., Kleypas, J.A., Farina, M., and Thompson, FlL., 2012. Rhodolith beds are major $CaCO_3$ bio-factories in the tropical south west Atlantic. PLoS one 7(4): e35171.

Anderson, C.A. 1950. Geology of Islands and Neighboring Land Areas. Part 1: 1940 E.W. Scripps Cruise to the Gulf of California. Geological Society of America Memoir 43, 53 p.

Applegate, S. P. 1978. Phyletic studies; part 1, Tiger sharks. Revista mexicana de ciencias geológicas, 2, 55-64.

Applegate, S.P. and Espinosa-Arrubarrena, L., 1981. The geology and selachian paleontology of Loma de Tirabuzón (Corkscrew Hill), Santa Rosalia, B.C.S. In: Ortleib, L. (ed.), Geology of northwestern México and southern Arizona. Instituto de Geología, Universidad Nacional Autónoma de México, México D.F.D., México, pp. 257–263.

Arango-Galván, C., Prol-Ledesma, R.M., and Torres-Vera, M.A. 2015. Geothermal prospects in the Baja California Peninsula. Geothermics 55: 39–57.

Aref, M.A.M., Basyoni, M.H., and Bachmann, G.H. 2014. Microbial and physical sedimentary structures in modern evaporitic coastal environments of Saudi Arabia and Egypt. Facies 60: 371–388.

Ashby, J.R., Ku, T.L., and Minch, J.A. 1987. Uranium series ages of corals from the upper Pleistocene Mulegé terrace, Baja California Sur, Mexico. Geology 15: 139–141.

Avila-Serrano, G.E., Téllez-Durarte, M.M., and Flessa, K.W. 2009. Ecological changes on the Colorado River delta – The shelly fauna evidence, pp. 95103. In: Johnson, M.E. and Ledesma-Vázquez, J. (eds.), Atlas of Coastal Ecosystems in the western Gulf of California, University of Arizona Press, Tucson, Arizona, 189 p.

Baarli, B.G., Santos, A., Silva, C.M. da, Ledesma-Vázquez, J., Mayoral, E., Cachão, M., and Johnson, M.E. 2012. Diverse macroids and rhodoliths from the Upper Pleistocene of Baja California Sur, Mexico. Journal of Coastal Research 28: 296–305.

Backus, D.H. and Johnson, M.E. 2009. Sand dunes on peninsular and island shores in the Gulf of California, pp. 117–133. In: Johnson, M.E. and Ledesma-Vázquez, J. (eds.), Atlas of Coastal Ecosystems in the western Gulf of California, University of Arizona Press, Tucson, Arizona, 189 p.

Backus, D.H. and Johnson, M.E. 2014. Stromatolitic mats from an uplifted Pleistocene lagoon at Punta Chivato on the Gulf of California (Meico). Palaios 29: 460–466.

Backus, D.H., Johnson, M.E., and Riosmena-Rodríguez, R. 2012. Distribution, sediment source, and coastal erosion of fan-delta systems on Isla Cerralvo (Lower Gulf of California, Mexico). Journal of Coastal Research 28: 210–224.

Barnes, L.G. 1998. The sequence of fossil marine mammal assemblages in México. Avances en Investigatción, Paleontología Vertebrados, Publicación Especial 1, Universidad Autonma del Estado de Hidalgo, pp. 26–79.

Barragán, R.M., Birkle, P., Porugal, E., Arellano, V.M., and Alvarez, J. 2001. Geochemical survey of medium temperate geothermal resources from the Baja California Peninsula and Sonora, México. Journal Volcanology and Geothermal Research 110: 101–119.

Bigioggero, B., Capaldi, G., Chiesa, S., Montrasio, A., Vessoli, L., Zanchi, A., 1988. Post-subduction magmatism in the Gulf of California: The Isla Coronados, Baja California Sur, Mexico. Instituto Lombardo (Rendiconti Scienze) B121: 117–132.

Brierley, C.M., Fedorov, A.V., Liu, Z., Herbert, T.D., Lawrence, K.T., and LaRiviere, J.P. 2009. Greatly expanded tropical warm pool and weakened Hadley circulation in the early Pliocene. Science 323: 1714–1718.

Bruno, D.E., Crowley, B.E., Gutak, J.M., Moroni, A., Nazarenko, O.V., Oheim, K.B., Ruban, D.A., Tiess, G., and Zorina, S.O. 2014. Paleogeography as geological heritage: Developing geosite classification. Earth-Science Reviews 138: 300–312.

Brusca, R.C. (ed). 2010. The Gulf of California: Biodiversity and Conservation. University of Arizona Press and Arizona-Sonora Desert Museum, Tucson, Arizona, 354 p.

Brusca, R.C. and Hendrickx, M.E. 2010. Invertebrate biodiversity and conservation in the Gulf of California, pp. 72–95. In: Brusca, R.C. (ed.), The Gulf of California, Biodiversity and Conservation. University of Arizona Press and Arizona-Sonora Desert Museum, Tucson, Arizona, 354 p.

Canet, C., Prol-Ledesma, R. M., Proenza, J. A., Rubio-Ramos, M. A., Forrest, M. J., Torres-Vera, M. A., and Rodríguez-Díaz, A. A. 2005. Mn–Ba–Hg mineralization at shallow submarine hydrothermal vents in Bahía Concepción, Baja California Sur, Mexico. Chemical Geology 224: 96–112.

Carriquiry, J.D. and Sanchez, A. 1999. Sedimentation in the Colorado River delta and Upper Gulf of California after nearly a century of discharge loss. Marine Geology 158: 125–145.

Carreño, A.L. and Helenes, J. 2002. Geology and ages of the islands, pp. 14-40. In: Case, T.J., Cody, M.L., and Ezcurra, E. (eds.), A New Island Biogeography of the Sea of Cortés, Oxford University Press, UK, 669 p.

Cintra-Buenrostro, C.E., Foster, M.S., and Meldahl, K.H., 2002. Response of nearshore marine assemblages to global change: a comparison of molluscan assemblages in Pleistocene and modern rhodolith beds in the southwestern Gulf of California, Mexico. Palaeogeographu, {Palaeoclimatologu, Palaeoecology 183: 229–230.

Crofts, R. and Gordon, J. E. 2015. Geoconservation in protected areas, pp. 531–568. In: G. L. Worboys, M. Lockwood, A. Kothari, S. Feary and I. Pulsford (eds), Protected Area Governance and Management, ANU Press, Canberra, Australia.

DeDiego-Forbis, T., Gorsline, D.R., Nava-Sanchez, E.N., Mack, L., and Banner, J., 2004. Late Pleistocene (Last Interglacial) terrace deposits, Bahía Coyote, Baja California Sur, Mexico. Quaternary International 120: 29–40.

Del Rio-Salas, R., Ochoa-Ladín, L., Eastoe, C.J., Ruiz. J.. Meza-Figueroa. D.. Valencia-Moreno, M., Zúñiga-Hernández, H., Zúñiga-Hernández, L., Moreno-Rodríguez, V., and Mendívil-Quijada, H. 2013. Genesis of manganese oxide mineralization in the Boleo region and Concepción , Baja California Sur: constraints from Pb-Sr isotopes and REE geochemistry. Revista Mexicana de Ciencias Geológicas 30: 482–499.

Des Marais, D.J. 2003. Biogeochemistry of hypersaline microbial mats illustrates the dynamics of modern microbial ecosystems and their early evolution of the biosphere. The Biological Bulletin 204: 160–167.

Dorsey, R.J. and Kidwell, S.M. 1999. Mixed carbonate-siliciclastic sedimentation on a tectonically active margin: Example from the Pliocene of Baja California Sur, Mexico. Geology 27: 935–938.

Dorsey, R.J., Umhoefer, P.J., and Falk, P.D. 1997. Earthquake clustering inferred from Pliocene Gilbert-type fan deltas in the Loreto basin, Baja California Sur, Mexico. Geology 25: 679–682.

Dorsey, R.J., Umhoefer, P.J., Ingle, J.C., Mayer, L. 2001. Late Miocene to Pliocene stratigraphic evolution of northeast Carmen Island, Gulf of California: implicaltions for oblique-rifting tectnics. Sedimentary Geology 144: 97–123.

Durham, J.W. 1950. Megascopic paleontology and marine stratigraphy. 1940 E.W. Scripps Cruise to the Gulf of California. Geological Society of America Memoir 43 (Part 2), 216 p.

Ellison, A.M., Farnsworth, E.J., and Merkt, R.E. 1999. Origins of mangrove ecosystems and the mangrove biodiversity anomaly. Global Ecology and Biogeography 8: 95–115.

Emhoff, K.F., Johnson, M.E., Backus, D.H, and Ledesma-Vázquez, J. 2012. Pliocene stratigraphy at Paredones Blancos: Significance of a massive crushed-rhodolith deposit on Isla Cerralvo, Baja California Sur (Mexico). Journal of Coastal Research 28: 234–243.

Eros, J.M., Johnson, M.E., and Backus, D.H. 2006. Rocky shores and development of the Pliocene-Pleistoene Arroyo Blanco Basin on Isla Carmen in the Gulf of California, Mexico. Canadian Journal of Earth Sciences 43: 1149–1164.

Estradas-Romero, A., Prol-Ledesma, R. M., and Zamudio-Resendiz, M. E. 2009. Relación de las características geoquímicas de fluidos hidrotermales con la abundancia y riqueza de especies del fitoplancton de Bahía Concepción, Baja California Sur, México. Boletín de la Sociedad Geológica Mexicana, 61: 8796.

Felix-Pico, E.F., Ramierez-Rodrigues, M., and Lopez-Rocha, J.A. 2015. Secondary production of bivalve populations in the mangrove swamps, p. 27–46. In: Riosmena-Rodríguez, R., Gonzáqlez-Acosta, A.F., and Muñiz-Salazar (eds.), The Arid Mangrove Forest from Baja California Peninsula, Volume 1, Nova Publishers, New York, 167 p.

Fierstine, H.L., Applegate, S.P., González-Barba, G., Schwennicke, T., and Espinosa-Arrubarrena, L. 2001. A fossil Blue Marlin (*Makaira nigricans* Lacépéde) from the Middle Facies of the Trindad Formation (Upper Miocene to Upper Pliocene), San José del Cabo Basin, Baja California Sur, México. Bulletin Southern California Academy of Science 107: 45-56.

Forrest, M. J., Ledesma-Vázquez, J., Ussler, W., Kulongoski, J. T., Hilton, D. R., and Greene, H. G. 2005. Gas geochemistry of a shallow submarine hydrothermal vent associated with the El Requesón fault zone, Bahía Concepción, Baja California Sur, México. Chemical Geology, 224: 82–95.

Forrest, M.J., and Ledesma-Vázquez, J. 2009. Active geothermal springs and Pliocene-Pleistocene examples, pp. 145–1155. In: Johnson, M.E. and Ledesma-Vázquez, J. (eds.), Atlas of Coastal Ecosystems in the western Gulf of California, University of Arizona Press, Tucson, Arizona, 189 p.

Foster, A.B. 1979. Environmental variation in a fossil scleractinian coral. Lethaia 12: 245–264.

Foster, M.S. 2001. Rhodoliths: Between rocks and soft places. Journal of Phycology 37: 659–667.

Foster, M.S., McConnico, L.M., Lundsten, L., Wadsworth, T., Kimball, T., Brooks, L.B. Medina-López, M., Risomena-Rodrígues, R., Hernández-Carmona, G., Vásquez-Elizondo, Johnson, S., and Steller, D.L. 2007. Diversity and natural history of a *Lithothamnion muelleri-Sargassum horridum* community in the Gulf of California. Ciencas Marinas 33: 367–384.

Gastil, R.G., Phillips, R.P., and Allison, E.C. 1971. Reconnaissance geologic map of the State of Baja California. Geological Society of America Memoir 140, Map sheet C.

Gedan, K.B., Kirwan, M.E., Wolanski, E., Barbier, E.B., and Silliman, B.R. 2011. The present and future role of coastal wetland vegetation in protecting shorelines: Answering recent challenges to the paradigm Climatic Change 106: 7–29.

Gonggrijp, G.P. 2000. Planning and management for geoconservation, pp. 2946. In: Barettino, D., Wimbledon, W.A.P., and Gallego, E. (eds.), Geological Heritage: Its Conservation and Management (Proceedings of Third International Symposium ProGEO on the Conservation of the Geological Heritage), Grafistaff, S.L., Madrid, Spain, 221 p.

González-Acosta, A.F., Rabadan-Sotelo, J.A., Ruiz-Campos, G.R., Del Moral-Fores, L.F., and Borges-Suza, J.M. 2015. A systematic list of fishes from an inshlar mangrove ecosystem in the Gulf of California, pp. 81–92. In: Riosmena-Rodriguez, R., Gonzáqlez-Acosta, A.F., and Muñiz-Salazar (eds.), The Arid Mangrove Forest from Baja California Peninsula, Volume 1, Nova Publishers, New York, 167 p.

Halfar,J., Eisele, M., Riegl, B., Hetzinger, S., Godinez-Orta, L. 2012. Modern rhodolith-dominated carbonates at Punta Chivato, Mexico. Geodiversitas 34: 99–113.

Hastings, P.H. Findley, L.T., and Van der Heiden, A.M. 2010. Fishes of the Gulf of California. In: Brusca, R.C. (ed.). The Gulf of California Biodiversity and Conservation. University of Arizona Press and Arizona-Sonora Desert Museum, Tucson, Arizona, pp. 96–118.

Hendrickx, M.E., Brusca, R.C., and Findley, L.T. 2005. A Distributional Checklist of the Macrofauna of the Gulf of California, Mexico. Part 1. Invertebrates. Arizona-Sonora Desert Museum, Tucson, Arizona 429 p.

Hetzinger, J., Halfar, J., Rioegl, B., and Godinez-Orta, L., 2006. Sedimentology and acustic mapping of modern rhodolith beds on a non-tropical carbonate shelf (Gulf of California, Mexico). Journal of Sedimentary Research 76: 670–682.

Holt, J.W., Holt, E.W., and Stock, J.M. 2000. An age constraint on Gulf of California rifting from the Santa Rosalía basin, Baja California Sur, Mexico. Geological Society of America Bulletin 112: 540–549.

Horodyski, R.J., Bloeser, B., and Vonder Haar, S. 1977. Laminated algal mats from a coastal lagoon, Laguana Mormona, Baja California, Mexico: Journal of Sedimentary Petrology 47: 680–696.

Ives, R.I. 1959. Shell dunes of the Sonoran shore. American Journal of Science 257: 449–457.

Johnson, M.E. 2002a. Discovering the Geology of Baja California – Six Hikes on the Southern Gulf Coast. University of Arizona Press, Tucson, Arizona 220 p.

Johnson, M.E. 2002b. Paleoislands in the stream: paleogeography and expected circulation patterns. Mémoire Special 24, Geobios 35: 96–106.

Johnson, M.E. 2014. Off-trail Adventures in Baja California – Exploring Landscapes and Geology on Gulf Shores and Islands. University of Arizona Press, Tucson, Arizona, 236 p.

Johnson, M.E., Baarli, B.G., Cachão, M., Silva, C.M., Ledesma-Vázquez, J., Mayoral, E.J., Ramalho, R.S., and Santos, A. 2012. Rhodoliths, uniformitarianism, and Darwin: Pleistocene and Recent carbonate deposits in the Cape Verde and Canary archipelagos. Palaeogeography, Palaeoclimatology, Palaeoecology 329–330: 83–100.

Johnson, M.E. and Backus, D.H. 2009. Beach deflation and accrual of Pliocene-Pleistocene coastal dunes of the Gulf of California region, pp. 134–144. In: Johnson, M.E. and

Ledesma-Vázquez, J. (eds.), Atlas of Coastal Ecosystems in the western Gulf of California, University of Arizona Press, Tucson, Arizona, 189 p.

Johnson, M.E., Backus, D.H., and Ledesma-Vázquez, J. 2003. Offset of Pliocene ramp facies at El Mangle by El Coloradito fault, Baja California Sur: Implications for transtensional tectonics, pp. 407-420. In: Johnson, S.E., Paterson, S.R., Fletcher, J.M., Girty, G.H., Kimbrough, D.L, and Martin-Baranjas, A. (eds.), Tectonic evolution of northwestern México and the southwestern USA. Geological Society of America Secial Paper 374, 478 p.

Johnson, M.E., Backus, D.H., and Riosmena-Rodríguez, R. 2009. Contribution of rhodoliths to the generation of Pliocene-Pleistoene limestone in the Gulf of California, pp. 83–94. In: Johnson, M.E. and Ledesma-Vázquez, J. (eds.), Atlas of Coastal Ecosystems in the western Gulf of California, University of Arizona Press, Tucson, Arizona, 189 p.

Johnson, M.E. and Ledesma-Vázquez, J. (eds.) 2009. Atlas of Coastal Ecosystems in the western Gulf of California, University of Arizona Press, Tucson, Arizona, 189 p.

Johnson, M.E., Ledesma-Vázquez, J., and Backus, D.H. 2016. Tectonic decapitation of a Pliocene mega-delta on Isla del Carmen in the Gulf of California (Mexico): And a river ran through it. Journal of Geology. In press.

Johnson, M.E., Ledesma-Vázquez, J., Backus, D.H., and González, M.R. 2012. Lagoon microbialites on Isla Angel de la Guarda and associated peninsular shores, Gulf of California (Mexico). Sedimentary Geology, 263–264: 76-84.

Johnson, M.E., Ledesma-Vázquez, J., Mayall, M.A., and Minch, J. 1997. Upper Pliocene stratigraphy and depositional systems: The Peninsula Concepción basins in Baja California Sur, Mexico, pp. 57-72. In: Johnson, M.E. and Ledesma-Vázquez, J. (eds.), Pliocene carbonates and related facies flanking the Gulf of California, Baja California, Mexico. Geological Society of America Special Paper 318, 171 p.

Johnson, M.E., Ledesma-Vázquez, J., and Montiel-Boehringe, A.Y. 2009. Growth of Pliocene-Pleistocene clam banks (Mollusca, Bivalvia) and trelated tectonic contraints in the Gulf of California, pp. 104-116. In: Johnson, M.E. and Ledesma-Vázquez, J. (eds.), Atlas of Coastal Ecosystems in the western Gulf of California, University of Arizona Press, Tucson, Arizona, 189 p.

Johnson, M.E., López-Pérez, R.A., Ransom, C.R., and Ledesma-Vázquez, J. 2007. Late Pleistocene coral-reef development on Isla Coronados, Gulf of California. Ciencias Marinas 33: 105–120.

Johnson, M.E., Perez, D.M., and Baarli, G.B. 2012. Rhodolith stranding event on a Pliocene rocky shore from Isla Cerralvo in the Lower Gulf of California (Mexico). Journal of Coastal Research 28: 225–233.

Kirkland, D.W., Bradbury, J.P., and Dean, W.E. Jr. 1966. Origin of Carmen Island salt deposit Baja California, Mexico. Journal of Geology 74: 932–938.

Kluesner, J., Lonsdale, P., and González-Fernández, A. 2014. Late Pleistocene cyclicity of sedimentation and spreading-center structure in the central Gulf of California. Marine Geology 347: 58–68.

Ledesma-Vázquez, J., Berry, R.W., Johnson, M.E., and Gutiérrez-Sanchez, S. 1997. El Mono chert: A shallow-water chert from the Pliocene Infierno Formation, Baja California Sur, Mexico, pp. 73-81. In: Johnson, M.E. and Ledesma-Vázquez, J. (eds.), Pliocene carbonates and related facies flanking the Gulf of California, Baja California, Mexico. Geological Society of America Special Paper 318, 171 p.

Ledesma-Vázquez, J., Carreño, A.L., Staines-Urias, F., and Johnson, M.E. 2006. The San Nicolás Formation: A proto-gulf extensional-related new lithostratigraphic unit at Bahía San Nicolás, Baja California Sur, Mexico. Journal of Coastal Research 22: 801–811.

Ledesma-Vázquez, J., Carreño, A.L., and Guardads-France, R. 2012. Biogenic coastal deposits: Isla del Carmen, Gulf of California, Mexico. Facies 58: 169–178.

Ledesma-Vázquez, J., Montiel-Boehringer, A.Y., Backus, D., Johnson, M., and Ferández-Díaz, V.Z. 2007. Armored mud balls in tidal environments: Pliocene in the Gulf of California. In Díaz-Martinez, E. and Rábano, I. (eds), Abstracts for the 4th European Meeting on the Palaeontology and Stratigraphy of Latin America. Cuademos del Museo Geominero, no 8. Madrid, Instituto Geológico y Minero de España, pp. 235-238.

Ledesma-Vázquez, J, Johnson, M.E., Backus, D.H., and Mirabal-Davila, C. 2007. Coastal evolution from transgressive barrier deposit to marine terrace on Isla Coronados, Baja California Sur, Mexico. Ciencias Marinas 33: 335–351.

Libbey, L.K. and Johnson, M.E. 1997. Upper Pleistocene rocky shores and intertidal biotas at Playa La Palmita (Baja California Sur, Mexico). Journal of Coastal Research 13: 216–225.

Liebig, P.M., Taylor, T.A. and Flessa, K.W. 2003. Bones on the beach: Marine mammal taphonomuy opf the Colorado Delta, Mexico. Palaios 18: 168–175.

Logan, B.W. 1961. Cryptozoon and associated stromatolites from the Recent, Shark Bay, Western Australia. Journal of Geology 69: 517–533.

López-Pérez, R. A. 2008. Fossil corals from the Gulf of California, México: Still a depauperate fauna but it bears more species than previously thought. Proceedings of the California Academy of Science, 4th series, 59: 515-531.

López-Pérez, R.A. 2012. Late Miocene to Pleistocene reef corals in the Gulf of California. Bulletins of American Paleontology 383: 1–78.

Margulis, L. and Sagan, D. 1987. Micro-Cosmos, Four Billion Years of Microbial Evolution. Allen & Unwin, London, 301 p.

Mark, C., Gupta, S., Carter, A., Mark, D.F., Gautheron, C., and Martin, A. 2014. Rift flank uplift at the Gulf of California: No requirement for asthenospheric upwelling. Geology 42: 259–262.

Marlett, C.M., 2014. Shells on a Desert Shore: Mollusks in the Seri World. University of Arizona Press, Tucson, Arizona, 480 pp.

Marrack, E.C. 1999. The relationship between water motion and living rhodolith beds in the southeastern Gulf of California, Mexico. Palaios 14: 159–171.

Martínez-Gutiérrez and Myer, L. 2004. Huracanes en Baja California, Mexico, y sus implicaciones en la sedimentación en El Golfo de California. Geos 24: 57–64.

Medina, E. 1999. Mangrove physiology: the challenge of salt, heat, and light stress under recurrent flooding, p. 109–126. In: Yáñez-Arancibia and Lara-Domínguez (eds.), Ecosistemas de Manglar en América Tropical. Intituto de Ecología A.C. México, UICN/ORM, Costa Rica, NOAA/NMFS, Silver Spring, MD, USA, 380 p.

Miranda-Avilés, R., Beraldi-Campesi, H., Puy-Alguiza, M.J., and Carreño, A.L. 2005. Estramatolitos, tufas y travertinos de la sección El Morro: Depósitos relacionados con la primera incursón marina en la Cuenca de Santa Roslía, Baja California Sur. Revista Mexicana de Ciencias Geológicas 22: 148–158.

Mortimer, E. and Carrapa, B. 2007. Footwall drainage evolution and scarp retreat in response to increasing fault displacement: Loreto fault, Baja California Sur, Mexico. Geology 35: 651–654.

Norris, J.N. 2014. Marine algae of the northern Gulf of California II: Rhodophyta. Smithsonian Contributions to Botany Number 96, 555 p.

Olvera, M.C., Aceves, J.S., Rendón, C., Valiente, C., Acosta, M.L.L., Rodríguez, B. 2004. La política ambiental Mexicana y la conservación del ambiente en Baja California Sur. Gaceta Ecológic 70, 45–56.

O'Neil, A. and O'Neil, D. 2001. Baja California: First Mission and Capital of Spanish California. Tio Press, Studio City, California, 282 p.

Pacheco-Ruiz, I., Zertuche-González, J.A., Meling-López, A.E., Riosmena-Rodríguez, R., and Orduña-Rojas, J. 2006. El límte norte de Rhizophora mangle L. en el golfo de California, México. Ciencia y Mar 28: 19–22.

Postma, G. 1995. Sea-level related architectural trends in coarse-grained delta complexes. Sedimentary Geology 98: 3–12.

Ramalho, R.S., Winckler, G., Madeira, J., Helffrich, G.R. Hipólito, A., Quartau, R., Adena, K., and Schaefer, J.M. 2015. Hazard potential of volcanic flank collapses raised by new megasunami evidence. Science Advances 1, e1500456.

Ravelo, A.C., Andreasen, D.H., Lyle, M., Lyle, A.O., and Wara, M.W. 2004. Regional climate shifts caused by gradual global cooling in the Pliocene epoch. Nature 429: 263–267.

Reyes-Bonilla, H. and Calderón-Aguilera, I..E. 1999. Population density, distribution and consumption reates of three corillivores at Cabo Pulmo reef, Gulf of California, Mexico. Marine Ecology 20: 347–357.

Reyes-Bonilla, H. and López-Pérez, R.A. 2009. Corals and coral-reef communities in the Gulf of California, pp. 45-

57. In: Johnson, M.E. and Ledesma-Váquez, J. (eds.), Atlas of Coastal Ecosystems in the western Gulf of California, University of Arizona Press, Tucson, Arizona.

Riosmena-Rodríguez, R., González-Acosta, A.F., and Muñiz-Salazar (eds.). 2015. The Arid Mangrove Forest from Baja California Peninsula. Volume 1, Nova Publishers, New York, 167 p.

Riosmena-Rodríguez, R., López-Calderón, J.M., Mariano-Meléndez, E., Sánchez-Rodrígues, A., and Fernández-Garcia, C. 2012. Size and distribution of rhodolith beds in the Loreto Marine Park: Their role in coastal processes. Journal of Coastal Research 28: 255–260.

Riosmena-Rodríguez, R., Steller, D.L., Hinojosa-Arango, G., and Foster, M.S. 2010. Reefs that rock and rool: Biology and conservation of rhodollith beds in the Gulf of California, p.49–71. In: Brusca, R. (ed.), The Gulf of Califlornia – Biodiversity and Conservation. Arizona-Sonora Desert Museum Studies in Natural History, University of Arizona Press, Tucson. Arizona, 354 p.

Romero-Vadillo, E., Zytsev, O., and Morales-Pérez, R. 2007. Tropical cyclone statistics in the northeastern Pacific. Atmóspfera 20: 197–213.

Ruban, D.A. 2010. Quantification of geodiversity and its loss. Proceedings of the Geologists' Association 121: 326–333.

Russell, P. and Johnson, M.E. 2000. Influence of seasonal winds on coastal carbonate dunes from the Recent and Plio-Pleistocene at Punta Chivato (Baja California Sur, Mexico). Journal of Coastal Research 16: 709–723.

Santamaria-del-Angel, E., Sebastiá-Frasquet, M.-T., Millán-Nuñez, R., González-Silvera, A., and Cajal-Medrano, R. 2015. Anthropocentric bias in management policies: Are we efficiently monitoring our ecosystems, pp. 2–11. In: Sebastiá-Frasquet, M.-T. (ed.). 2015. Coastal Ecosystems: Experiences and Recommendations for Enviornmental Monitoring Programs. Nova Publishers, New York, 219 p.

Santos, I.R., Lechuga-Deveze, C., Peterson, R.N., and Burnetts, W.C. 2011. Tracing submarine hydrothermal inputs into a coatral bay in Baja California using radon. Chemical Geology 282: 1–10.

Schlanger, S.O. and Johnston, C.J. 1969. Algal banks near La Paz, Baja California – Modern analogs of source areas of transported shallow-water fossil in pre-alpine flysche deposits. Palaeo 6: 1090–1098.

Seminoff, J.A. 2010. Sea turtles of the Gulf of California. In: Brusca, R.C. (ed.). The Gulf of California Biodiversity and Conservation. University of Arizona Press and Arizona-Sonora Desert Museum, Tucson, Arizona, pp. 135–167.

Sewell, A.A., Johnson, M.E., Backus, D.H., and Ledesma-Vázquez, J. 2007. Rhodolith detritus impounded by a coastal dune on Isla Coronados, Gulf of California. Ciencias Marinas 33: 483–494.

Shepard, F.P. 1950. Submarine topography of the Gulf of California. Part 3: 1940 E.W. Scripps Cruise to the Gulf of California. Geological Society of America Memoir 43, 32 p.

Simian, M.E. and Johnson, M.E. 1997. Development and foundering of the Pliocene Santa Ines Archipelago in the Gulf of California: Baja California Sur, Mexico, pp. 25–38. In: Johnson, M.E. and Ledesma-Vázqauez, J. (eds.), Pliocene carbonates and related facies flanikign the Gulf of California, Mexico. Geological Society of America Special Paper 318, 171 p.

Skudder, P.A. III, Backus, D.H., Goodwin, D.H., and Johnson, M.E. 2006. Sequestration of carbonate shell material in coastal dunes on the Gulf of California (Baja California Sur, Mexico). Journal of Coastal Research 22: 611–624.

Siqueiros-Beltrones, D.A. 2008. Role of pro-thrombolites in the geomorpholoty of a coastal lagoon. Pacific Science 62: 257–269.

Steinbeck, J. and Ricketts, E.F. 1941. Sea of Cortez: A Leisurely Journal of Travel and Research. Viking Press, New York, 598 p.

Steinbeck, J. 1951. The Log from the Sea of Cortez. Viking Press, New York, 288 p.

Steller, D.L. and Foster, M.S. 1995. Environmental factors influencing distribution and morphology of rhodoliths in Bahía Concepción, B.C.S., México. Journal of Experimental Marine Biology and Ecology 194: 201–212.

Tierney, P.W. and Johnson, M.E. 2012. Stabilization role of crustose corallilne algae during Late Pleistocene reef development on Isla Cerralvo, Baja California Sur (Mexico). Journal of Coastal Research 28: 244–254.

Vasconcelos, C., Dittrich;, M., and McKenzie, J.A. 2014. Evidence of microbiocoenosis in the formation of laminae in modern stromatolites. Facies 60: 3–13.

Wara, M.W., Ravelo, A.C., and Delaney, M.L. 2005. Permanent El Niño-like conditions during the Pliocene Warm Period. Science 309: 758–761.

Wilson, I. and Rocha, V. 1955. Geology and mineral deposit of the Boleo Copper District Baja California, México. U.S. Geological Survey Professional Paper 273: 37–39.

Wimbledon, W. A. P., Ishchenko, A. A., Gerasimenko, N. P., Karis, L. O., Suominen, V., Johansson, C. E., & Freden, C. 2000. Geosites—an IUGS initiative: science supported by conservation. Geological Heritage: its Conservation and Management. Madrid (Spain), 69-94.

Umhoefer, P.J., Schwennicke, T., Del Margo, M.T., Ruiz-Geraldo, G., Ingle, J.C., and McIntosh, W. 2007. Transtensional fault-termination basins: an important basin type illustrated by the Pliocene San José basin and related basins in the southern Gulf of California, Mexico. Basin Research 19: 297–322.